SCIENCE WORKSHOP

CLOCKS · SCALES & MEASUREMENTS

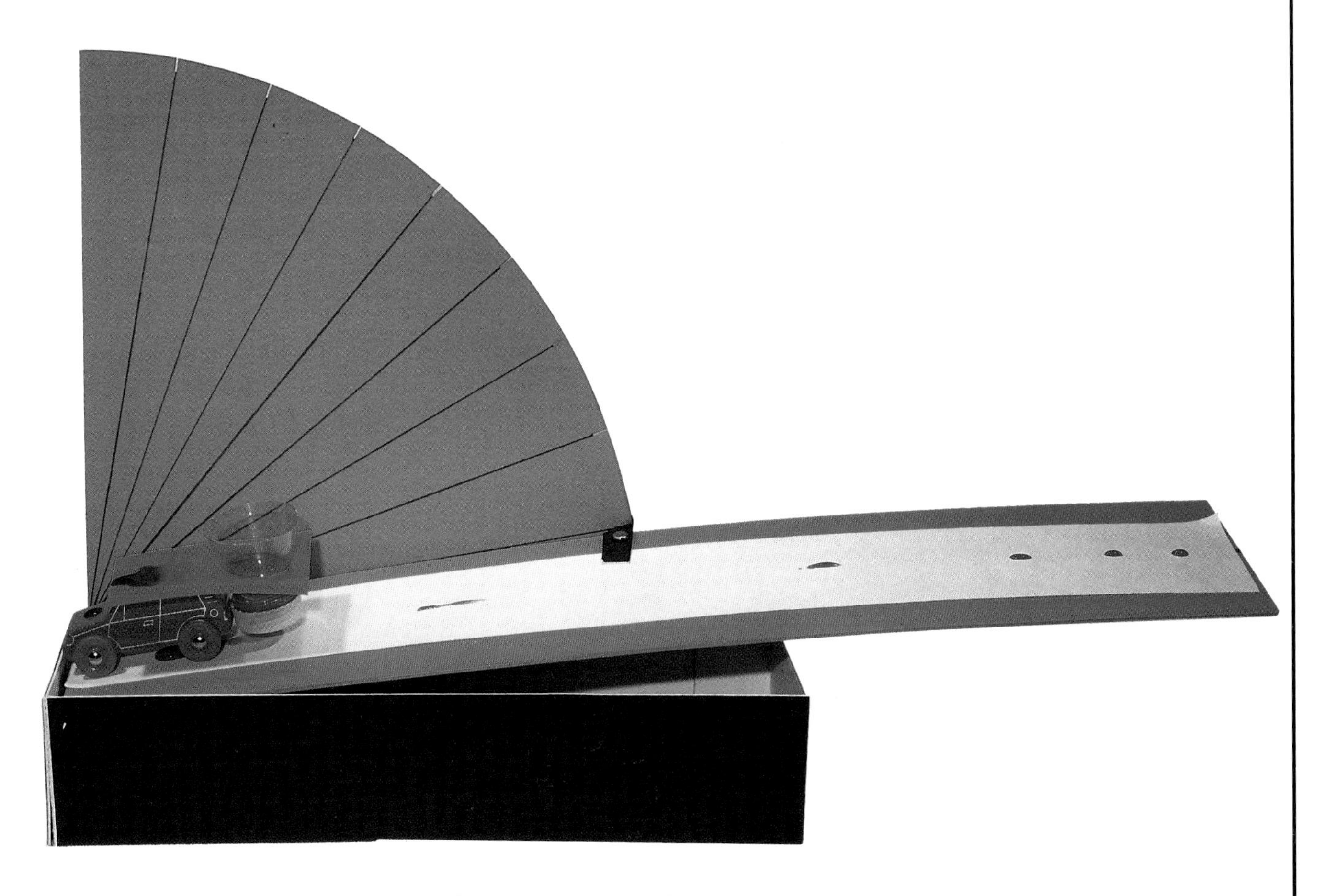

Pam Robson

A WATTS BOOK
LONDON NEW YORK SYDNEY

Design David West
Children's Book Design
Designer Steve Woosnam-Savage
Editor Suzanne Melia
Picture Researcher Emma Krikler
Illustrator Ian Thompson
Consultant Bryson Gore

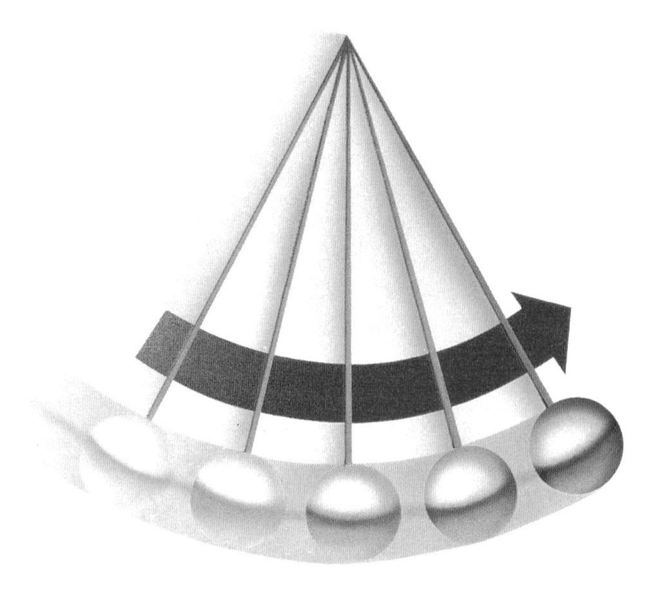

This edition produced in 1995

Created and designed by
N.W. Books
28 Percy Street
London W1P 0LD

First published in
Great Britain in 1993 by
Watts Books
96 Leonard Street
London EC2A 4RH

ISBN 0 7496 1174 X (Hardback)
0 7496 2081 1 (Paperback)

A CIP catalogue record for this book
is available from the British Library.

Printed in Belgium.

Contents

Photocredits
All the photographs in this book are by Roger Vlitos apart from pages; 6 top: Science Photo Library; 8 top, 22 top & 24 top: Eye Ubiquitous; 12 top right: Reproduced by permission of the Trustees of the Science Museum; 18 top & 20 top: Spectrum Colour Library.

INTRODUCTION

Listen out for the questions "How many?" and "How much?" During an average day, you are likely to hear them quite a few times. The answers to both questions use numbers. But, the first question is answered by counting and the second by measuring. We use measurements every day of our lives. Each time we look at a clock, we are reading a measurement of time. Whenever we travel, short or long distances, we need to know how long the journey will take, how far away our destination is and in which direction we are travelling. Buying food and clothes also involves measurement and a familiarity with the scales and units being used. Standard measuring tools and scales must be used around the world for the system to work. The imperial system (feet and inches) was developed around the 1200s, but most countries now use the metric system (metres and centimetres). A standard metre is one-ten millionth of a straight line distance from the North Pole to the Equator. If you think this sounds fascinating, read on to find out how scales and units of measurement were developed and how they are used. You'll realise what a chaotic world it would be without them!

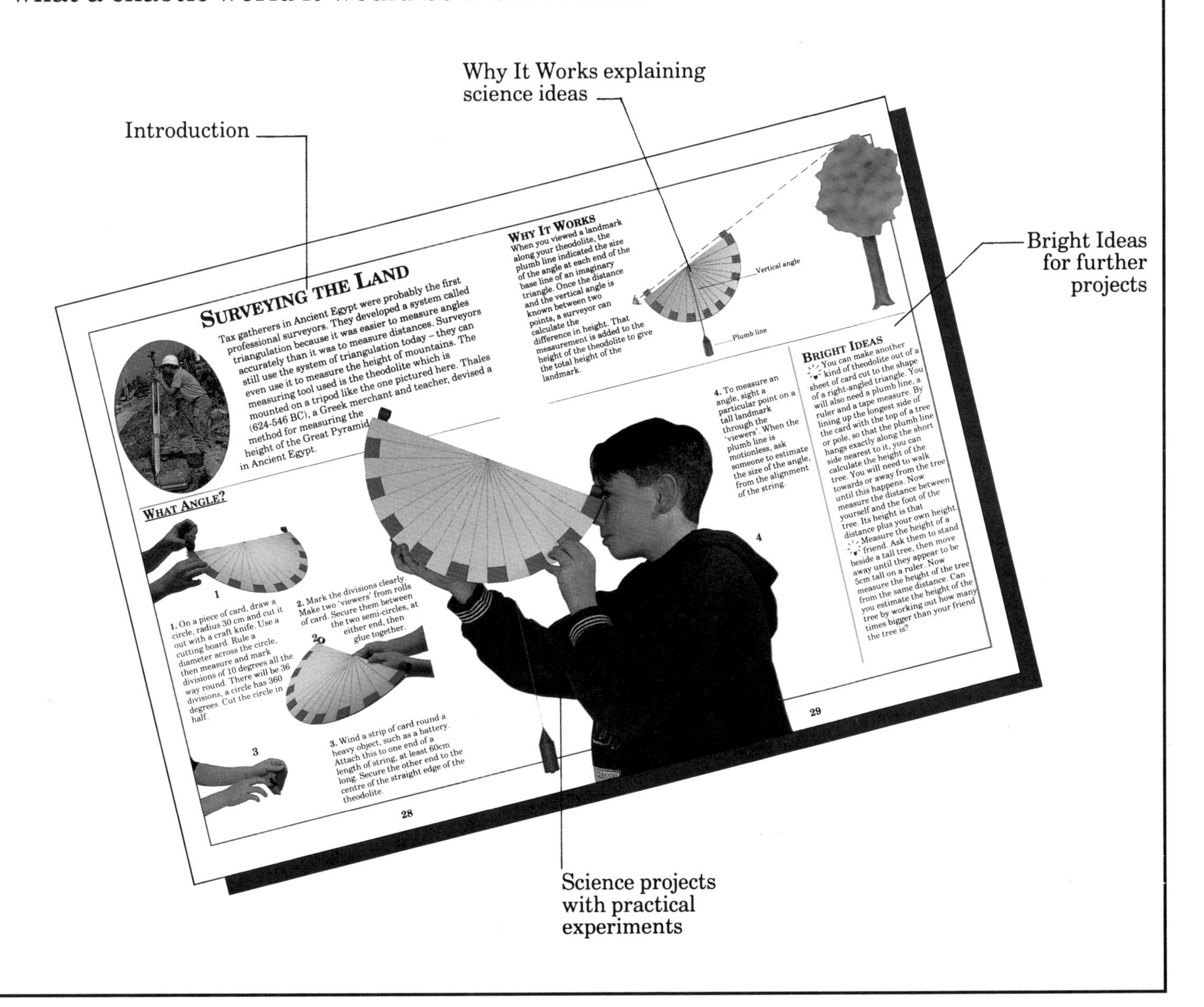

SURVEYING THE LAND

Tax gatherers in Ancient Egypt were probably the first professional surveyors. They developed a system called triangulation because it was easier to measure angles accurately than it was to measure distances. Surveyors still use the system of triangulation today – they can even use it to measure the height of mountains. The measuring tool used is the theodolite which is mounted on a tripod like the one pictured here. Thales (624-546 BC), a Greek merchant and teacher, devised a method for measuring the height of the Great Pyramid in Ancient Egypt.

WHAT ANGLE?

1. On a piece of card, draw a circle, radius 30 cm and cut it out with a craft knife. Use a cutting board. Rule a diameter across the circle, then measure and mark divisions of 10 degrees all the way round. There will be 36 divisions, a circle has 360 degrees. Cut the circle in half.

2. Mark the divisions clearly. Make two 'viewers' from rolls of card. Secure them between the two semi-circles, at either end, then glue together.

3. Wind a strip of card round a heavy object, such as a battery. Attach this to one end of a length of string, at least 60cm long. Secure the other end to the centre of the straight edge of the theodolite.

28

WHY IT WORKS

When you viewed a landmark along your theodolite, the plumb line indicated the size of the angle at each end of the base line of an imaginary triangle. Once the distance and the vertical angle is known between two points, a surveyor can calculate the difference in height. That measurement is added to the height of the theodolite to give the total height of the landmark.

4. To measure an angle, sight a particular point on a tall landmark through the 'viewers'. When the plumb line is motionless, ask someone to estimate the size of the angle from the alignment of the string.

BRIGHT IDEAS

You can make another kind of theodolite out of a sheet of card cut to the shape of a right-angled triangle. You will also need a plumb line, a ruler and a tape measure. By lining up the longest side of the card with the top of a tree or pole, so that the plumb line hangs exactly along the short side nearest to it, you can calculate the height of the tree. You will need to walk towards or away from the tree until this happens. Now measure the distance between yourself and the foot of the tree. Its height is that distance plus your own height.

Measure the height of a friend. Ask them to stand beside a tall tree, then move away until they appear to be 5cm tall on a ruler. Now measure the height of the tree from the same distance. Can you estimate the height of the tree by working out how many times bigger than your friend the tree is?

29

The Workshop

A science workshop is a place to test ideas, perform experiments and make discoveries. To prove many scientific facts, you don't need a lot of fancy equipment. In fact, everything you need for a basic workshop can be found around your home or school. Read through these pages, and then use your imagination to add to your "home laboratory". As you work your way through this book, you should enjoy completing the projects and seeing your models work. Remember, though, that from a scientific point of view, these projects are just the starting point. For example, when you finish making the egg-timer on pages 10/11, ask your own questions like "What would happen if I used sand?", "Would a larger hole let the sand flow faster?", and so on. Also by sharing ideas, you will learn more. Experimenting with equipment, as well as with ideas, will give you the most accurate results. In this way you will build up your workshop as you go along.

Making Models

Before you begin, read through all the steps. Then, make a list of the things you need and collect them together. Next, think about the project so that you have a clear idea of what you are about to do. Finally, take your time in putting the pieces together. You will find that your projects work best if you wait while glue or paint dries. If something goes wrong, retrace your steps. And, if you can't fix it, start over again. Every scientist makes mistakes, but the best ones know when to begin again!

Safety Warnings

Make sure that an adult knows what you are doing at all times. Cutting the top off a plastic bottle can be difficult and dangerous if you use sharp scissors. Ask an adult to do this for you. Always be careful with balloons and plastic bags. Never cover your face with them. If you spill any water, wipe it up right away. Slippery surfaces are dangerous. Clean up after you have finished.

General Tips

There are at least two parts to every experiment: experimenting with materials and testing a science "fact". If you don't have all the materials, experiment with others instead. For example, if you can't find any glass bottles, use some wine glasses instead. Once you've finished experimenting, read your notes thoroughly and think about what happened, evaluating your measurements and observations. What conclusions can you draw from your results?

Experimenting

Always conduct a "fair test". This means changing one thing at a time for each stage of an experiment. In this way, you can always tell which change caused a different result. As you go along, record what you see. Ask questions such as "why?", "how?" and "what if?". Then test your model and write down the answers you arrive at. Compare your results to those of your class-mates or friends.

CALENDARS

Before the invention of the clock, people had to rely on nature's timekeepers - the Sun, the Moon and the stars. The daily movement of the Sun across the sky provided the simplest unit, the solar day. The time period of a year was estimated by watching the seasons, and the constancy of the lunar cycle led to the division of each year into months. Traditionally, calendars were controlled by priests. They were devised either by counting days or by following the phases of the Moon. Nowadays, the Gregorian calendar is most common. This was worked out by Pope Gregory XIII in the 1580s.

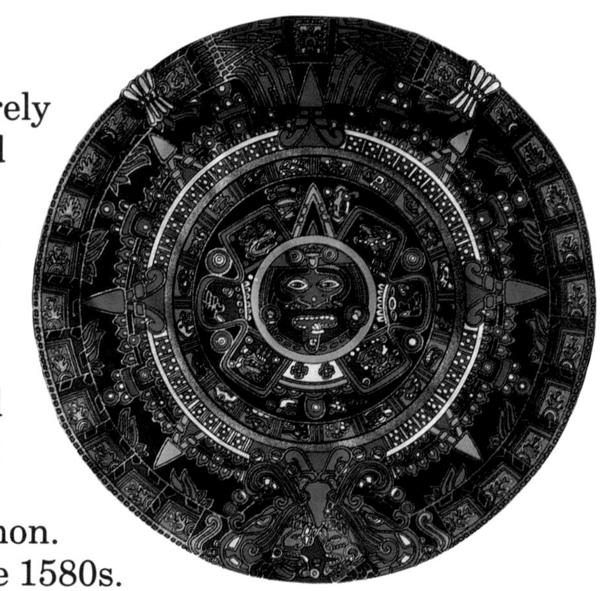

DAY BY DAY

1. Cut out two circles of card. The largest should be 30 cm across, the other 28 cm across. Stick one on top of another and divide into 12 equal pieces to indicate the months.

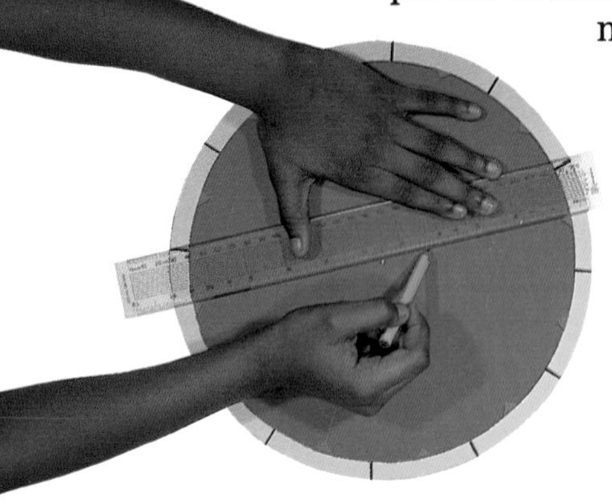

1

2

2. Cut out 12 paper circles of 1 cm across. Find out the number of days for each month and write them around the edge of each small circle. Now stick them in order around the large circle. The first day of the month should be nearest the edge as shown.

3. Cut out another card circle 27 cm across. Cut out a hole, radius 1cm, to correspond with the position of the paper circles. Cover the hole with stiff, transparent plastic. Attach a red arrow marker, as shown.

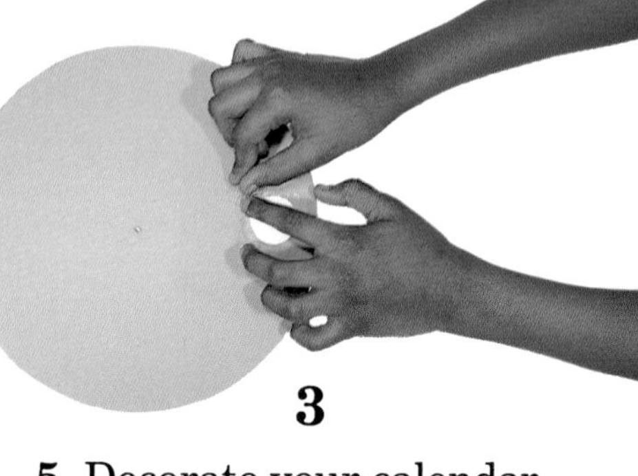

3

4. Cut out a card circle, radius 1cm. Carefully make a tiny 'window', to view the date through. Position it over the 1cm hole, and fix it to the plastic with a paper fastener, so it turns.

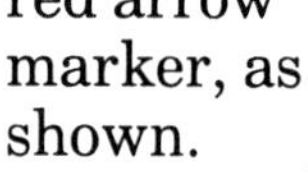

4

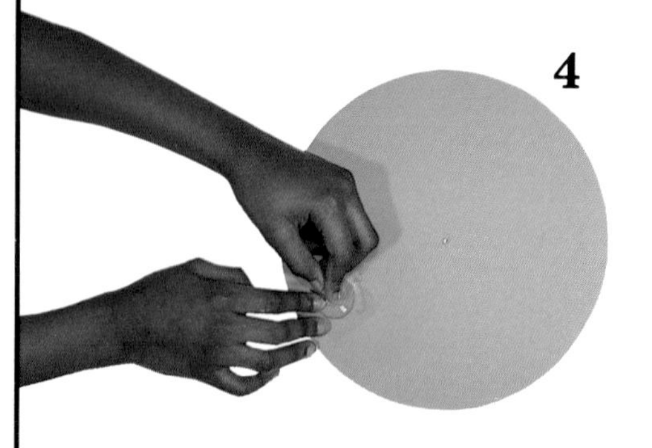

5

5. Decorate your calendar before joining the separate sections together. Position the smaller circle centrally over the larger circle and join them together with a split pin. Rotate your calendar until it is set on the correct day for the current month.

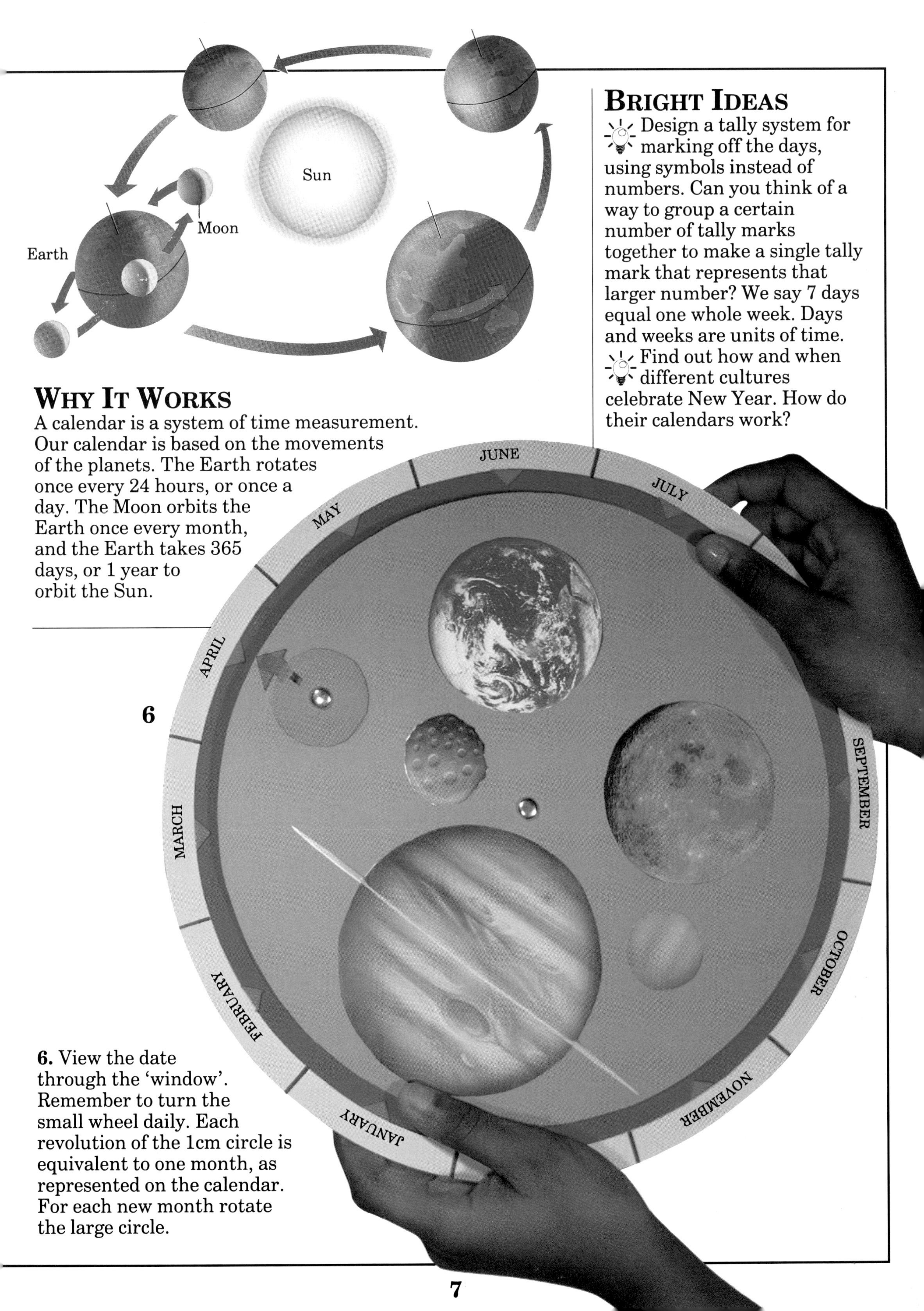

Bright Ideas

Design a tally system for marking off the days, using symbols instead of numbers. Can you think of a way to group a certain number of tally marks together to make a single tally mark that represents that larger number? We say 7 days equal one whole week. Days and weeks are units of time.

Find out how and when different cultures celebrate New Year. How do their calendars work?

Why It Works

A calendar is a system of time measurement. Our calendar is based on the movements of the planets. The Earth rotates once every 24 hours, or once a day. The Moon orbits the Earth once every month, and the Earth takes 365 days, or 1 year to orbit the Sun.

6. View the date through the 'window'. Remember to turn the small wheel daily. Each revolution of the 1cm circle is equivalent to one month, as represented on the calendar. For each new month rotate the large circle.

Plotting the Stars

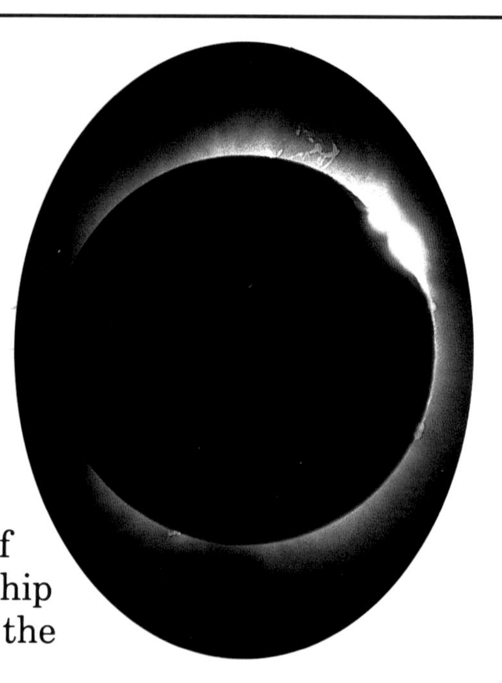

Astronomy is the study of the stars, planets and other objects in the universe. For centuries, astronomers have striven to learn more about our Univ erse. Through observation and careful measurement, using scientific tools like the telescope, we now know with certainty that the Sun is the centre of our solar system. Accurate measurement of star distance is a science developed over the centuries by astronomers like Tycho Brahe (1546-1601). Centuries ago, sailors calculated time at night by observing the movements of star clusters near the fixed Pole Star. Watches aboard ship were timed from the position of these constellations in the

Star Time

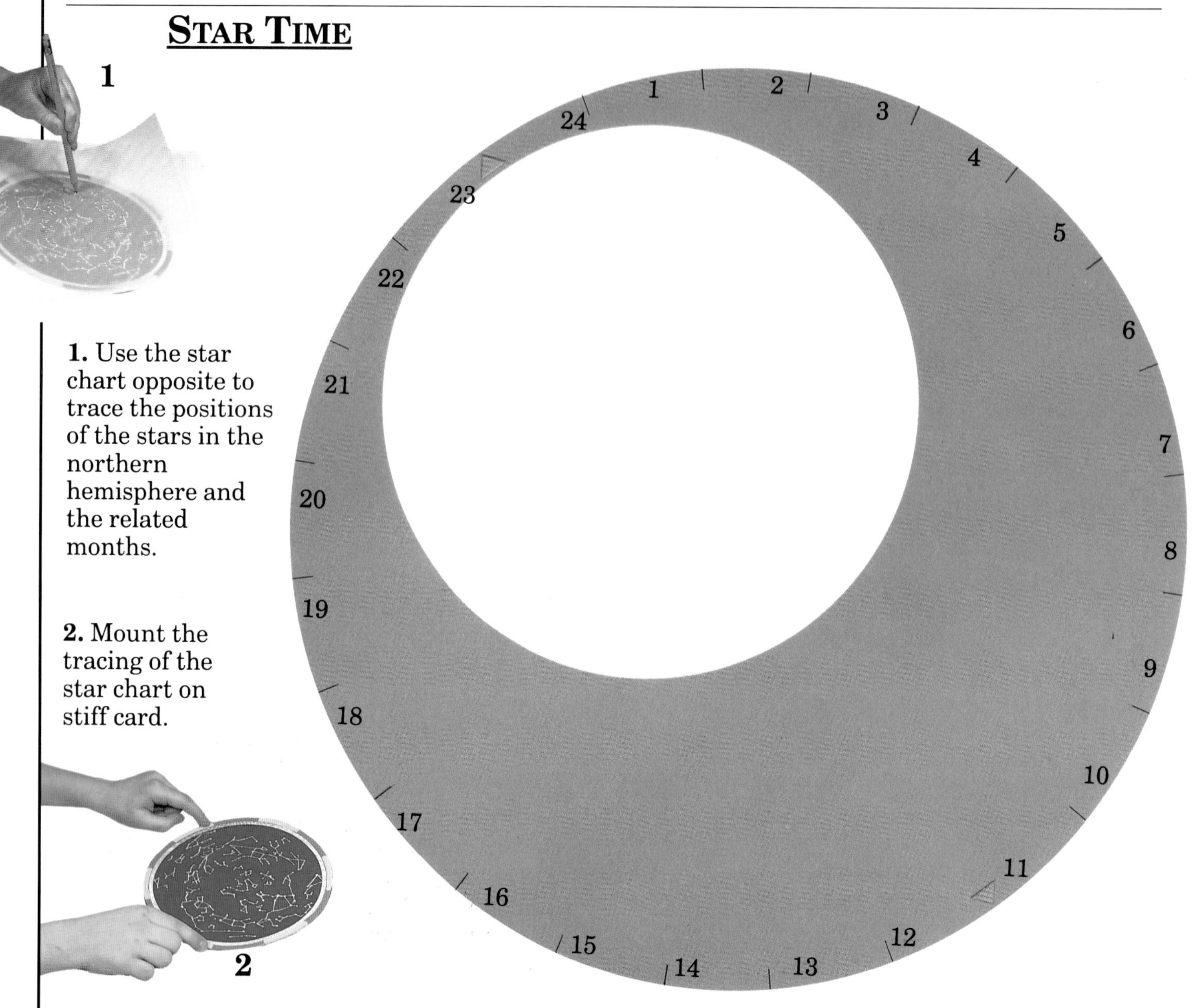

1. Use the star chart opposite to trace the positions of the stars in the northern hemisphere and the related months.

2. Mount the tracing of the star chart on stiff card.

3. Trace the shape on the opposite page onto card. Mark the 24 hours of the day, starting with Noon at the bottom

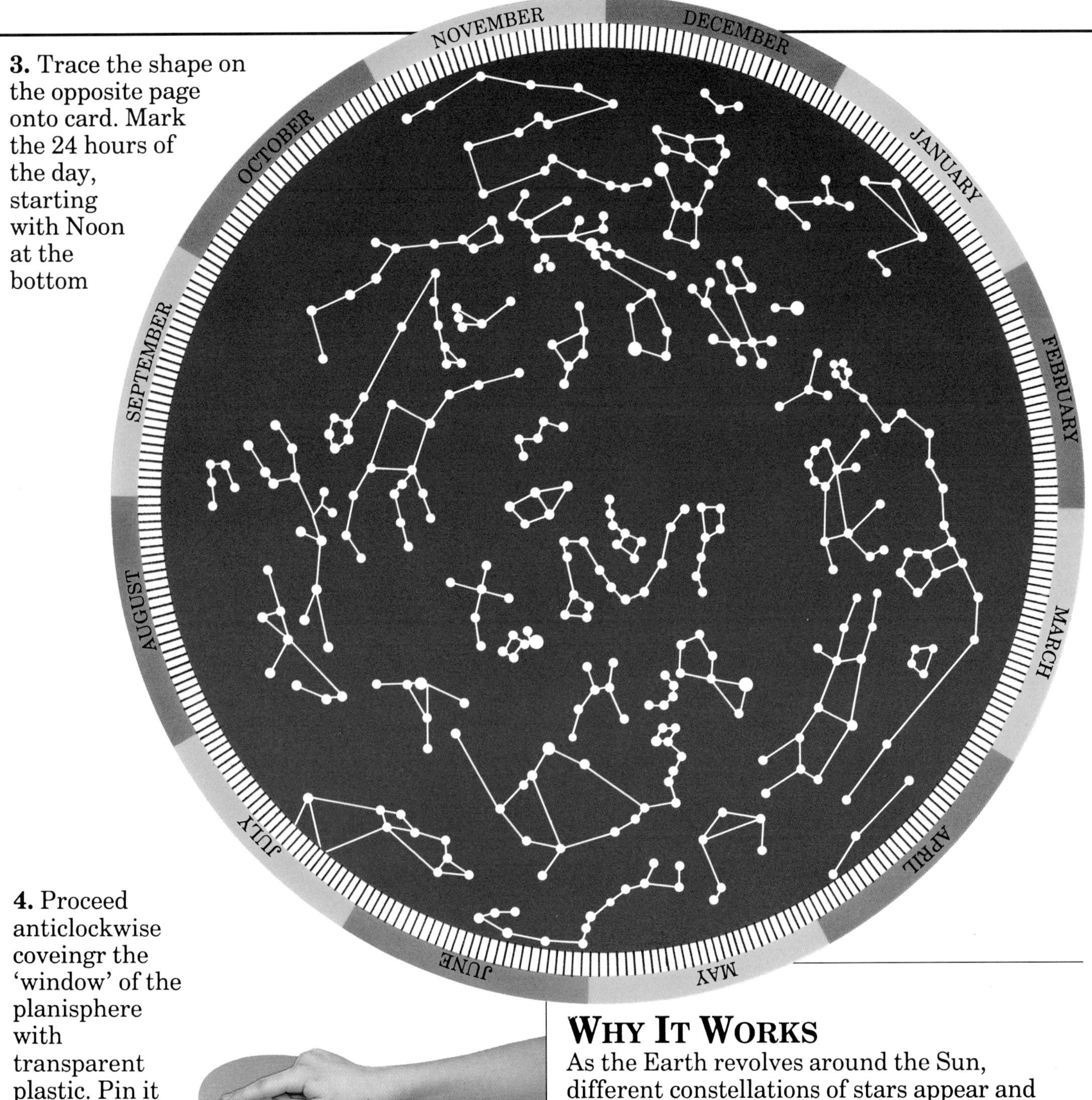

4. Proceed anticlockwise coveingr the 'window' of the planisphere with transparent plastic. Pin it to the starchart through the centre.

3

5. On a starry night, rotate the planisphere until the time of day, marked on the edge, is lined up with the appropriate month, at the bottom of the star chart. Compare what you see with the stars in the sky.

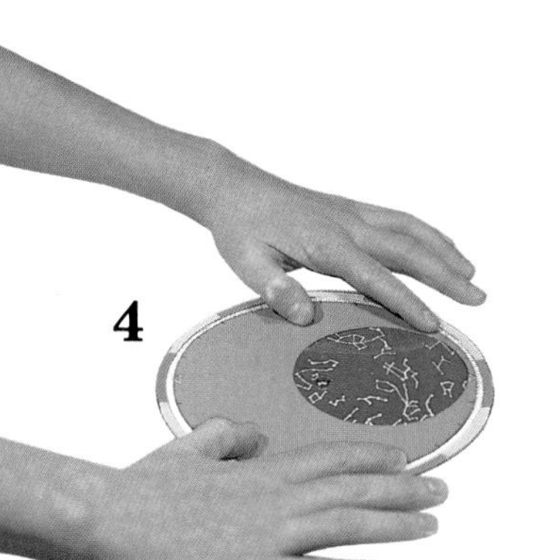

Why It Works

As the Earth revolves around the Sun, different constellations of stars appear and disappear, in a continous cycle that can be observed. This planisphere shows where these groups of stars can be seen at any given time of the year, in the northern hemisphere. By matching up the time of day with the date, you can view the stars that should be visible in the night sky through the cut out 'window'. The planisphere should be held up and viewed from underneath. The stars visible through the window should match those in the sky. The Sun is a star. It is the only star close enough to look like a ball. The other billions of stars are so far away, they appear to be pinpoints of light.

MEASURING ROTATION

The Earth revolves, on its own tilted axis, once every 24 hours. The Sun appears to rise and set. The rotation of the Earth was demonstrated in 1851, by J.B.L. Foucault, a French physicist. He realised that although a swinging pendulum seems to change direction throughout the day, it is really the Earth that rotates beneath it whilst the swing of the pendulum tries to remain constant. By studying the movement of the planets, astronomers have noted a gradual slowing of the Earth's rotation. They have calculated that the length of a day increases by about 1.7 milliseconds per year.

IN A SPIN

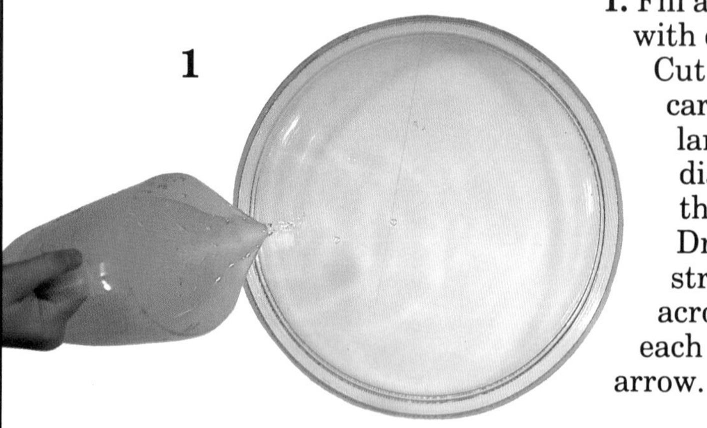

1

1. Fill a circular bowl with clean, cold water. Cut out a circle of card, slightly larger than the diameter of the bowl. Draw a straight line across, marking each end with an arrow.

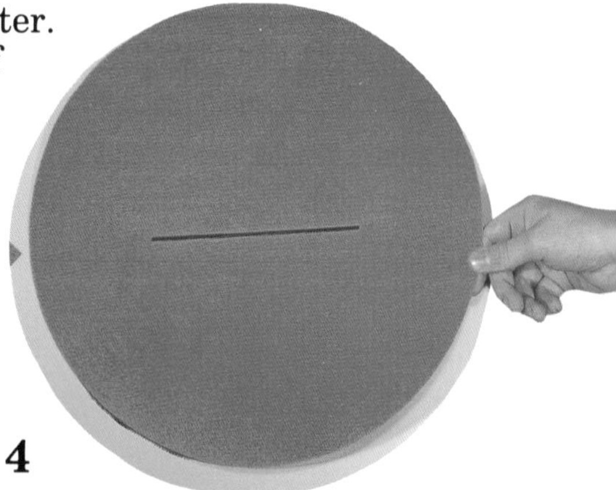

4

2. Find a flat surface, free of vibration. Position the card on the surface, then stand the bowl of water on top. Leave it to stand until the surface of the water is still.

3

3. Cut another circle of card, slightly larger than the bowl. Rule a line across the centre. Using a craft knife, on a board, cut out a slit approximately 2mm wide, as shown.

4. Rest the card circle on top of the bowl. Line up the slit with the marks on the card underneath the bowl, as shown. Prepare some powdered cork and keep in a dry container.

2

5. If the water is still, gently sprinkle the cork powder over the slit in the card. It will fall through to make a narrow line of powder on the surface of the water.

5

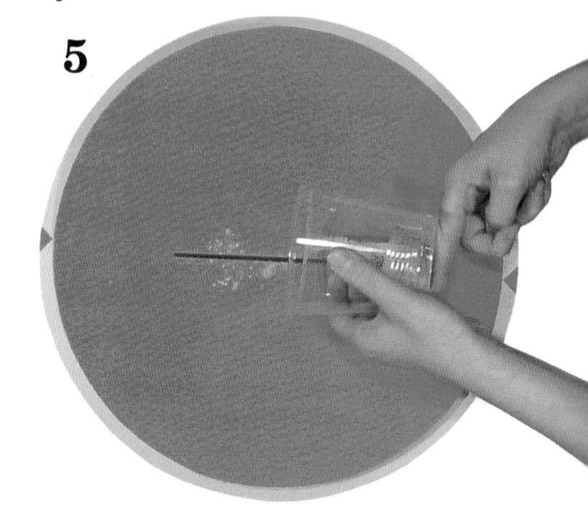

Why It Works

The line of cork dust appears to have rotated after the bowl has been standing for a number of hours. In reality, the Earth and the bowl have rotated beneath the line of dust, which remained stationary. This is because the Earth travels through 360 degrees each time a rotation is completed. Every place on Earth travels through 15 degrees each hour of the day. Artificial time zones have been created around the world to accomodate time differences.

Earth

Direction of rotation

Bright Ideas

Make a pendulum by pushing a needle right through an apple and suspending it inside an empty box with thread. The thread should be just the right length to let the needle sticking out of the apple to touch a plate underneath. Sprinkle some fine salt on the plate and start the pendulum swinging. Turn the plate slowly and watch the grooves form as the pendulum swings in the same plane and the plate rotates. The plate represents the Earth as it rotated in Foucault's pendulum experiment.

6. Do not touch the bowl. Leave it undisturbed for several hours. You will find, after some time, that the line of dust appears to have rotated and is no longer in line with the arrows.

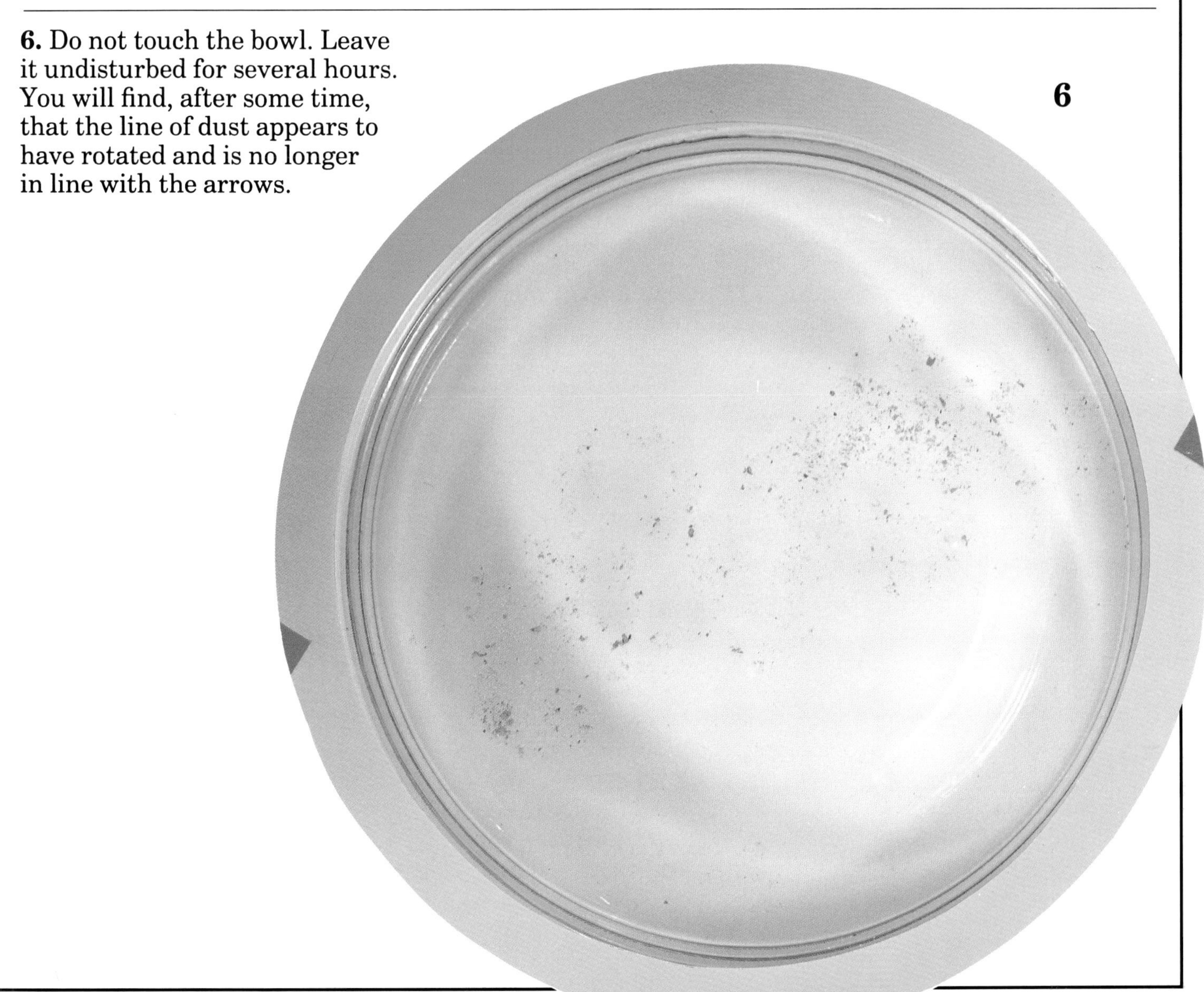

Estimating Time

Hours are artificial units of time, first introduced by the Ancient Egyptians when they observed that shadows follow a similar pattern of movement each day. Shadow clocks and sundials were early time-measurement tools. In the 14th century, the sandglass, or hourglass, was popular, but it could only be used to estimate periods of time, varying from minutes to hours. It could not indicate the time of day. On board ship, a four-hour glass timed 'watch' for the crew, until John Harrison invented the more accurate chronometer in 1735. This also calculated longitude and latitude. Early sandglasses were filled with powdered eggshell or marble dust.

Time is Running Out

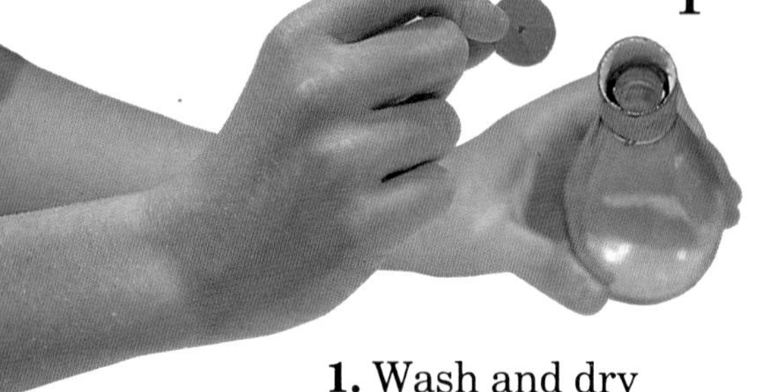

1

1. Wash and dry two small bottles thoroughly. Make an open-ended cylinder of card and slide it over the top of one bottle. Cut out a disc of card to fit inside and make a hole in its centre.

2

2. Make sure the other bottle is absolutely dry. Now, carefully pour a measured amount of salt into it.

3

3. Position the empty bottle on top of the salt-filled bottle by sliding the card cylinder over the neck of the lower bottle.

4

4. Check that the card 'seal' round the middle is secure. Carefully turn your timer over and observe what happens.

5. You can estimate the time taken for the salt to slide into the lower half of the timer by marking the side of the bottle with evenly spaced divisions. Use a stop watch to check exactly how long it takes for all the salt to slide to the bottom.

WHY IT WORKS

The upper vessel of the timer holds just enough salt to run through a hole, of a given size, in a given period of time. The force of gravity pulls the salt down, through the hole and into the bottom container. The salt grains must be absolutely dry, so they don't stick. The size of the hole between the two vessels will determine the speed at which the grains will flow, but once established, the rate of flow will not vary. The total period of time depends on the quantity of salt.

Salt

BRIGHT IDEAS

- Remove the regulator from between the two bottles and replace it with another, in which a hole of a different diameter has been made. Repeat this exercise a number of times, changing the size of the hole each time. What do you discover?
- Design a sandglass that will run for exactly 3 minutes - use it as an eggtimer when you boil an egg.
- Can you design another kind of sandglass that runs for a much longer period of time? Shape a funnel from card, and insert it into the neck of a measuring container. Fill the funnel with sand and time how long it takes to pour through into the container below. Standardise your method of reading the scale - the top of the mound of sand will be concave.

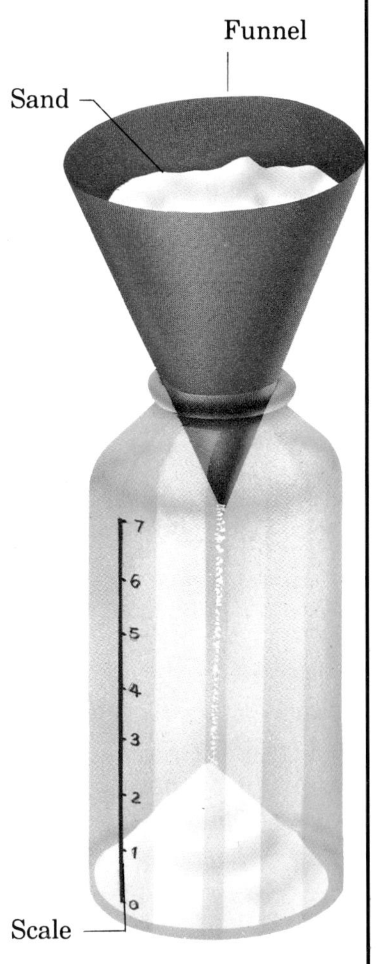

Measuring Time

The accurate measurement of small units of time was not possible until 1583. The Italian physicist, Galileo (1564-1642) observed the swinging of a lamp in Pisa cathedral, and by timing its motion against his own pulse, he concluded that a free pendulum was isochronous - every swing, wide or narrow, took the same time. The Dutch scientist, Christiaan Huygens (1629-1695), constructed the first mechanical clock with a pendulum mechanism in 1657. In 1921, W. H. Shortt invented a pendulum clock so accurate that it was used in the Royal Observatory in Greenwich. The measurement standard for time is the atomic clock. These clocks measure time so accurately, they will not gain or lose more than a second in 300 years.

Tick Tock

1

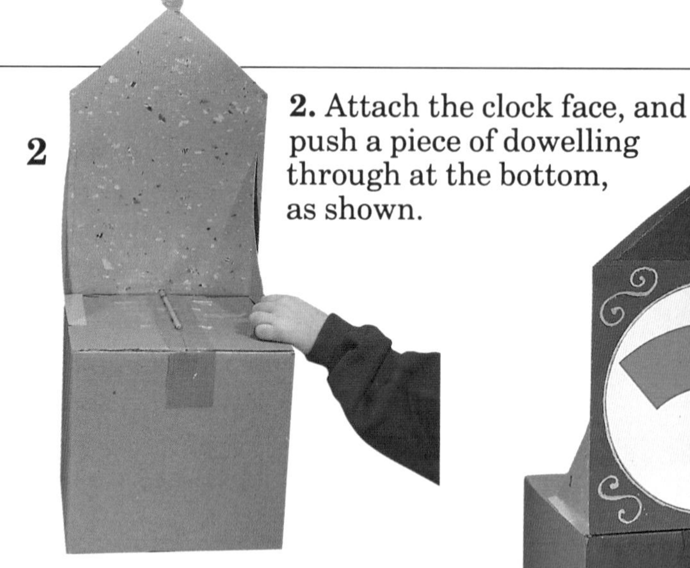

2

2. Attach the clock face, and push a piece of dowelling through at the bottom, as shown.

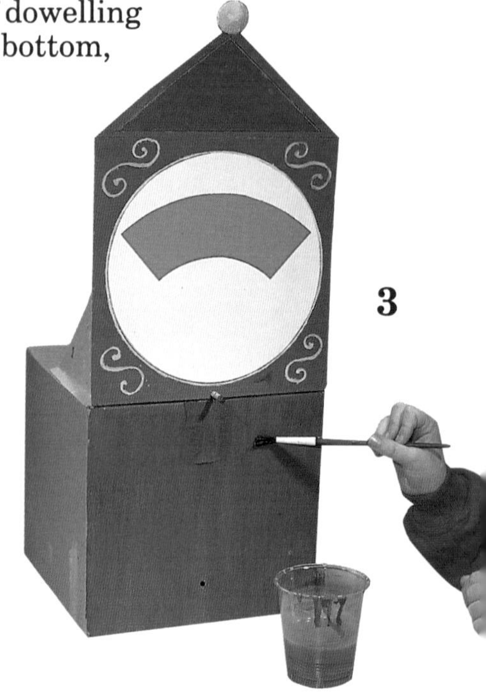

3

1. Cut the top off a large, square box, as shown. This will be the base of the clock. Cut a second piece of card, the same width as the box, to be the clock face.

3. Decorate your clock face and design and draw a 'dial'. Mark the indicator scale when you can observe the size of the pendulum swing.

4. The pendulum can be any length you wish. Use dowelling with a circle of card attached. To give weight to the bob, fix plasticine behind the card.

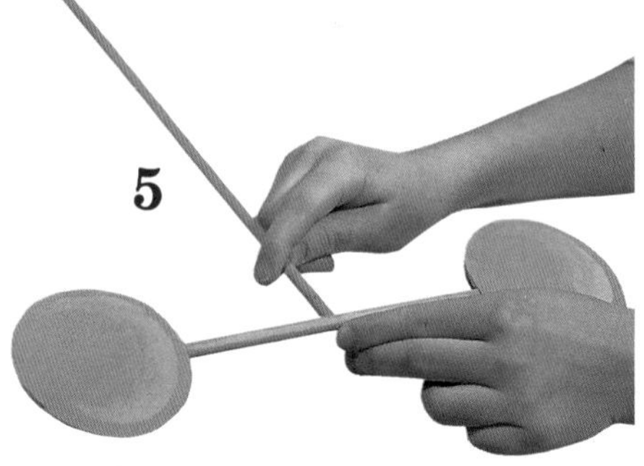

5

6. Make an arrow to be the 'hand' that points to the dial. Pin the pendulum to the piece of dowelling so it can swing freely.

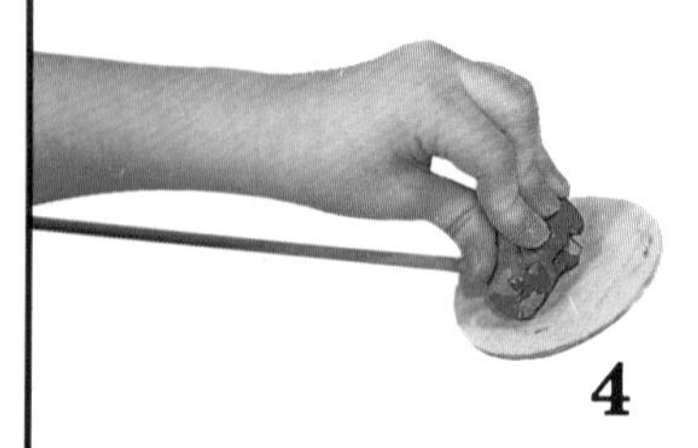

4

5. Balance the pendulum by fixing a short length of dowelling, with a circle of card at each end, across the top, as shown.

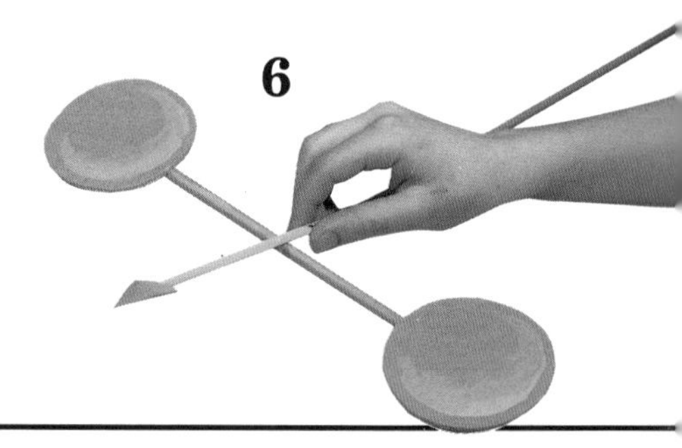

6

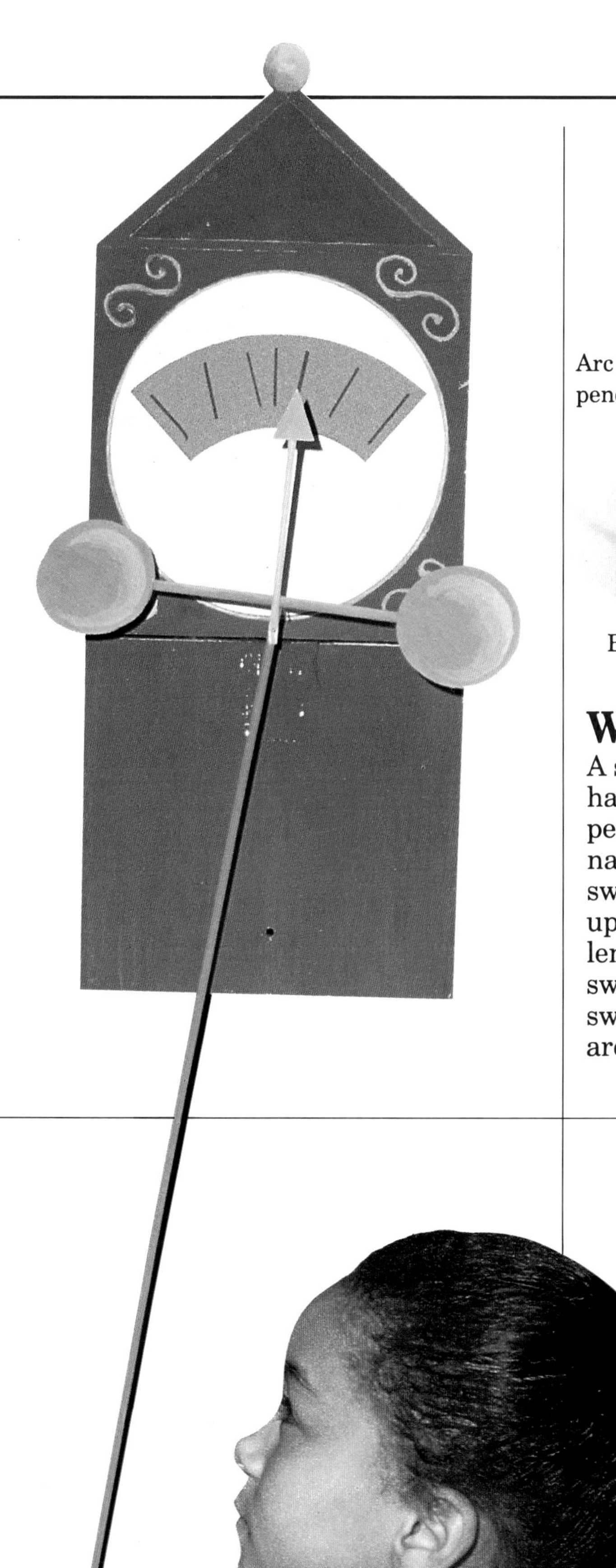

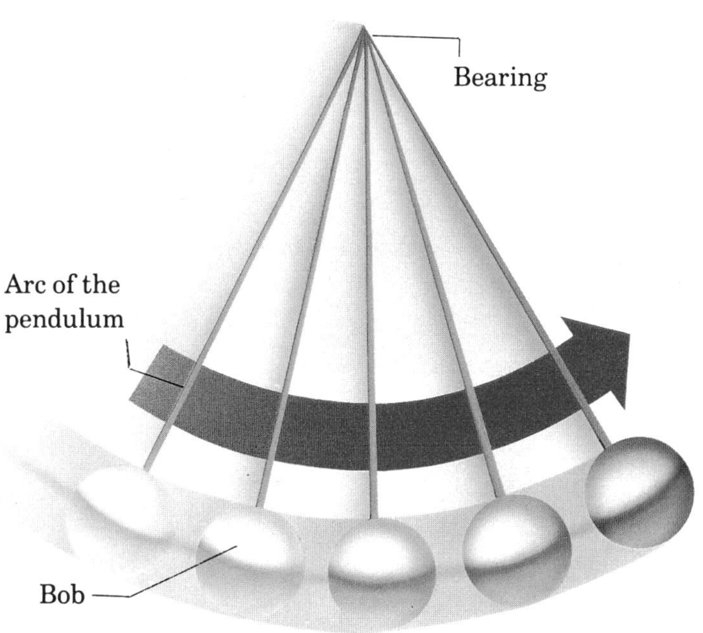

Why It Works

A simple pendulum consists of a weight hanging at the end of a string or wire. When a pendulum swings to and fro, it follows its own natural frequency due to gravity, and always swings at the same rate. This movement splits up time into small measureable units. The length of the pendulum effects the rate of swing. The shorter the length, the faster the swing. The time it takes to swing through one arc is called the period of oscillation.

Bright Ideas

- Alter the size of the bob on the pendulum. What difference, if any, does this make?
- What happens if you change the height of the starting point? Use squared paper mounted behind the pendulum to mark each starting point accurately.
- Can you design and make a pendulum that takes exactly one second for each swing? Try lengthening or shortening the string until you get it right. Try it out by counting 60 swings - it should take exactly 60 seconds.

Regulating Time

In 1808, Sir William Congreve designed the rolling ball clock. The timing mechanism was a ball that took exactly 15 seconds to roll, in a zig zag fashion, in grooves along an inclined ramp. An accurate pocket watch was only possible after Robert Hooke discovered, in 1685, that a vibrating spring could provide a regular rhythm for a watch. The modern watch may use the natural vibrations of quartz crystal, 100,000 times per second, to keep time.

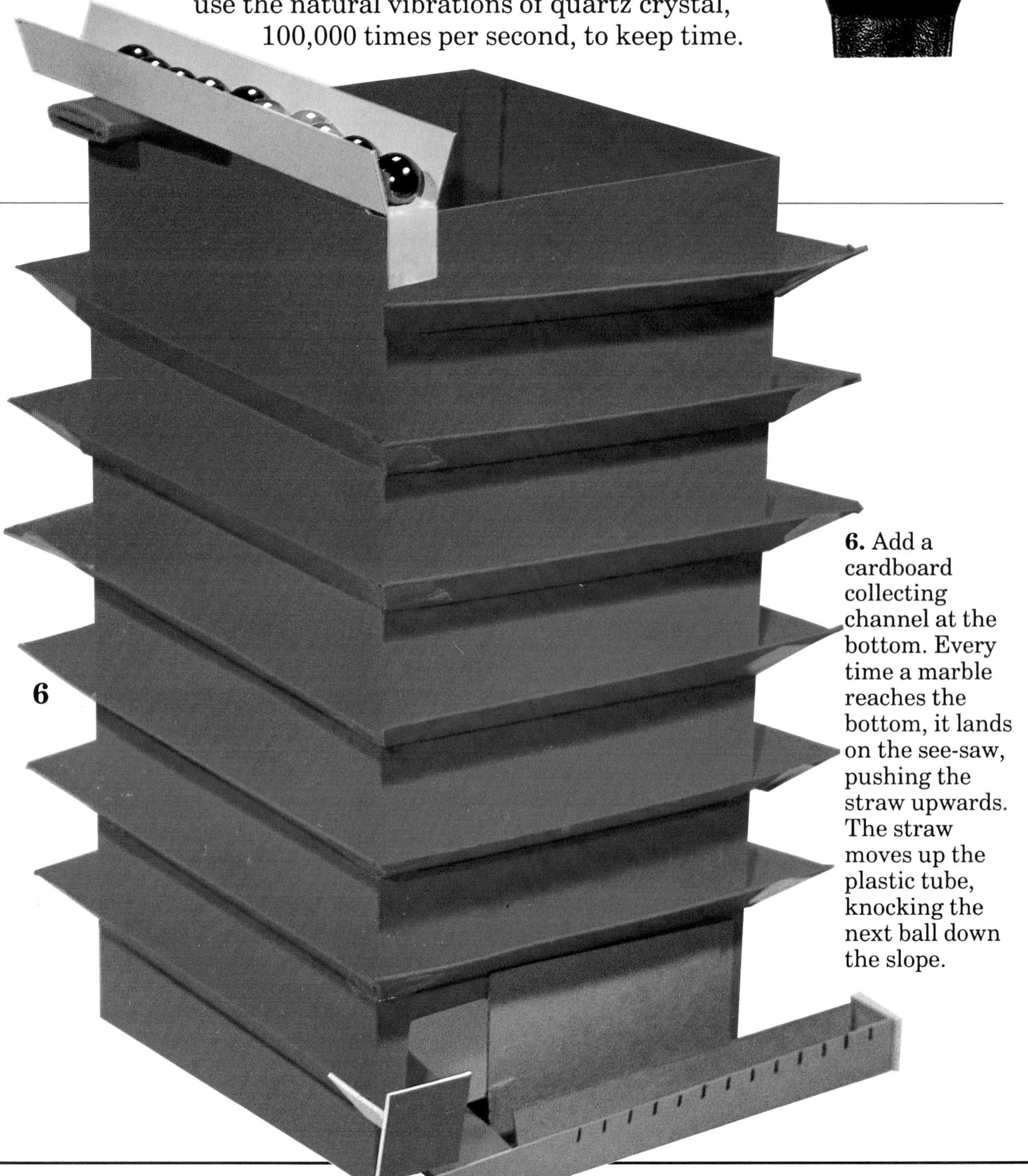

6. Add a cardboard collecting channel at the bottom. Every time a marble reaches the bottom, it lands on the see-saw, pushing the straw upwards. The straw moves up the plastic tube, knocking the next ball down the slope.

HELTER SKELTER

1

1. Take a rectangle of card, 15cm x 36cm, and divide into four columns. Cut 6 parallel slots in each column, sloping slightly from left to right.

2

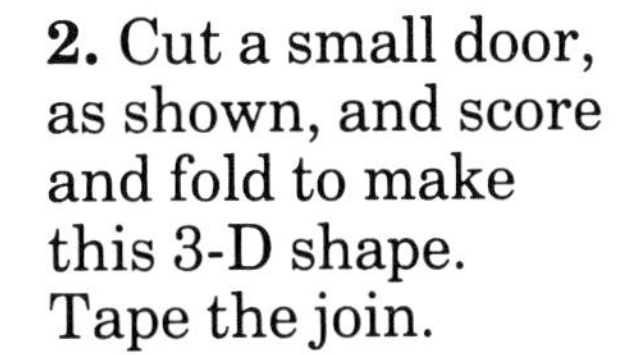

2. Cut a small door, as shown, and score and fold to make this 3-D shape. Tape the join.

3. Now measure, cut and score 24 card slots to fit into each of the slits. Slot them into position to make a zig-zag channel for the marbles to roll along.

4

4. Make a channel for the marbles from card. Cut a marble-sized hole at one end.

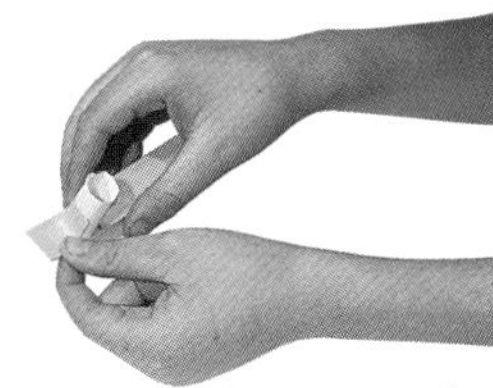

5

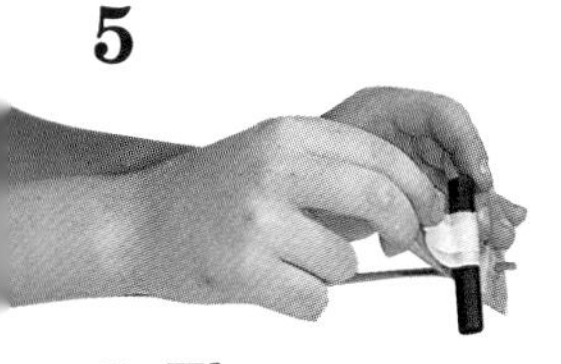

5. Tape a narrow plastic tube inside. Glue in place, raising it up at one end.

6. The see-saw mechanism at the bottom is a piece of card, weighted underneath. A piece of tubing acts as the pivot and a straw runs from one end, up into the plastic tube above.

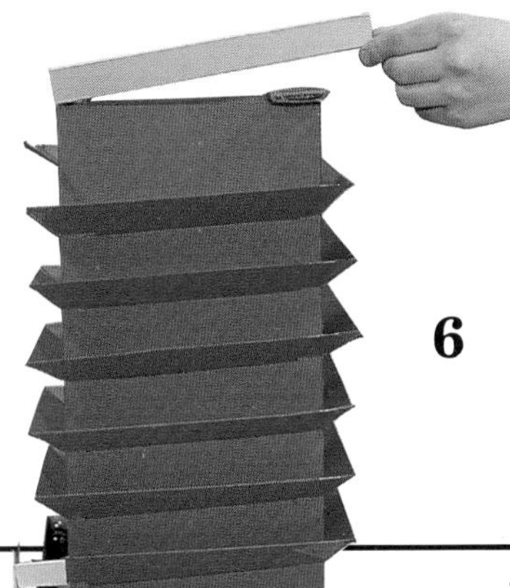

6

WHY IT WORKS

The rolling marble (1) is pulled down the slope by gravity. The force of gravity causes the marbles to accelerate as they fall. The marbles slow down when they are forced to change direction. Each marble takes exactly the same amount of time to roll down the slope. If this time is known, it can be multiplied by the number of marbles at the bottom (2) to calculate how much time has passed.

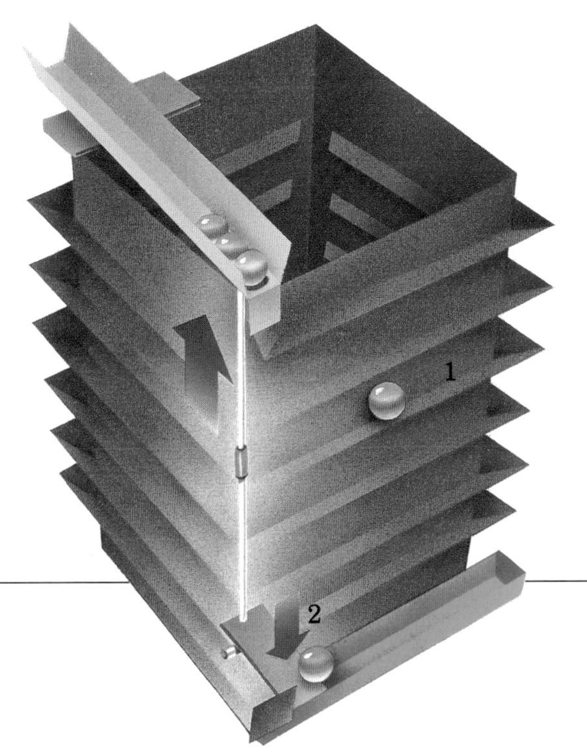

BRIGHT IDEAS

What happens if you use marbles of varying size? Is it possible to time each marble's descent with total accuracy? Does each take exactly the same amount of time? What is the total time taken for all of the marbles to reach the bottom?

Design and make a water clock like the one shown here. Will the flow be constant? How does the water clock compare with the sandglass? Can you design a water clock that takes exactly 10 minutes to empty?

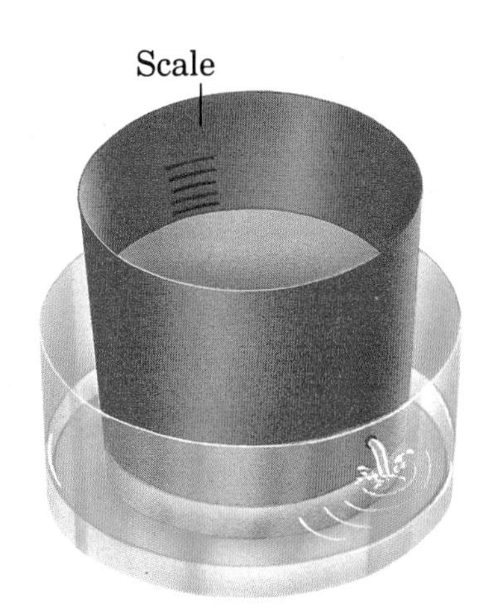

SPEED AND ACCELERATION

Speed is how fast something is moving. Velocity describes both speed and direction. When a car turns a corner, speed may stay the same, but velocity changes. Speed describes how far an object travels in a period of time. For example, a snail moves at about 0.05 kilometres per hour, while Concorde travels about 2,200 kilometres per hour. A speedometer, like the one pictured here, indicates how fast a car is moving. Acceleration is how much the speed increases in a period of time. A decrease in speed is called deceleration.

AT FULL SPEED

1

1. Cut the "road" from stiff card as wide as the shoe box, but twice as long. Secure a small peice of card in the middle of one side, with a split pin.

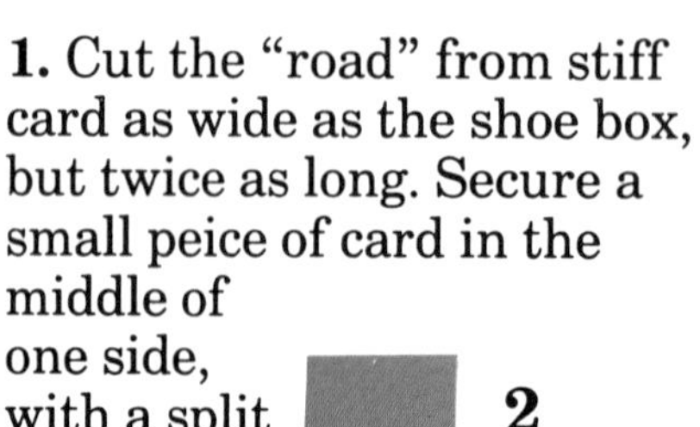

2

2. Tape one end of the road to the narrow end of the shoe box. It can be lifted to different heights.

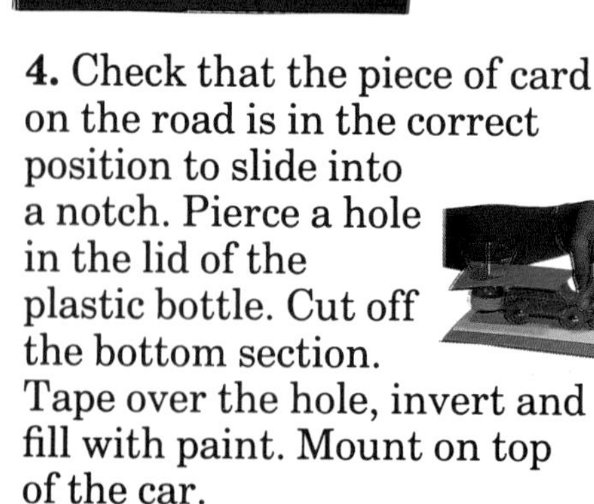

3

3. Cut out one quarter of a circle. Divide it into angles of 10 degrees, and cut slits along the edge. Attach to the side of the box.

4. Check that the piece of card on the road is in the correct position to slide into a notch. Pierce a hole in the lid of the plastic bottle. Cut off the bottom section. Tape over the hole, invert and fill with paint. Mount on top of the car.

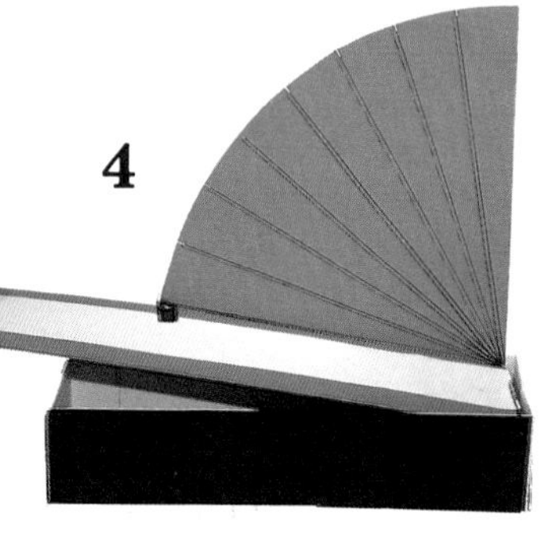

4

WHY IT WORKS

By observing the distance between the drops of paint on the inclined ramps you can estimate the speed of the vehicle. When they are close together, the speed is slowest. If the spaces are uniform, the vehicle must be travelling at a constant speed. If the spaces widen, the vehicle has accelerated, if they narrow it has decelerated. The bigger the mass of an object, the greater the force needed to make it move. When the slope is steeper, the truck accelerates faster. The spaces between the drops are wider apart towards the bottom of the slope. The speed can be calculated by dividing the distance by the time taken.

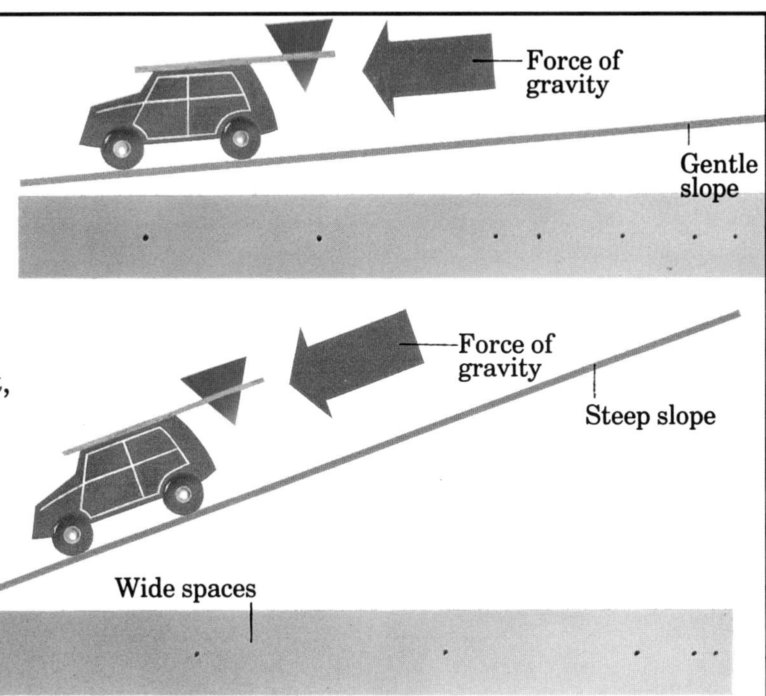

BRIGHT IDEAS

Use a variety of toy cars on a sloping ramp and experiment with differing gradients. Which car travels furthest? Did you use the same 'push' each time to ensure a fair test?

Run the same car down a variety of gradients, and allow it to run into a shoe box each time. Measure the distance that the box has moved. The distance that it moves depends on the speed.

Try running on a beach at various speeds. Use a stop-watch to time yourself and measure the spaces between the footprints. Your footprints will be further apart, the faster you run.

5

5. Set the gradient of the road. Place a long piece of paper over the road to record your results each time. Position the car at the top of the slope. Remove the tape just before it is released. Time each run accurately. Be careful not to push the car.

WEIGHT AND GRAVITY

In space, objects are weightless. This is because weight is due to the forces of gravity. Gravity is the force of attraction that acts towards the Earth's core, pulling us all downwards. Sir Isaac Newton (1642-1727) developed the theory of gravitational pull. The Earth is held in orbit around the Sun by the pull of gravity. Gravitational pull is visible daily in the movement of the tides. They are a direct consequence of the gravitational pull between Earth, Moon and Sun. The weight of an object can vary slightly at different places on Earth.

PULL HARD!

1

1. Measure out a vertical indicating scale on a long sheet of card. Space the divisions evenly.

2

2. Cut the base from a plastic bottle. Paint it a bright colour. This will be a weighing pan. Pierce 2 holes directly opposite each other in the sides and attach a piece of thread, as shown.

3

3. Use the painted top of a plastic bottle as a holder, from which to suspend the balance. Cut a flap, as shown here, and make two holes in it. These will be the screw holes through which it can be attached to a length of wood.

4. Tie an elastic band to the thread already attached to the weighing pan. Loop the free end to the plastic holder. When suspended the weighing pan must be level. If it is not, adjust the elastic.

4

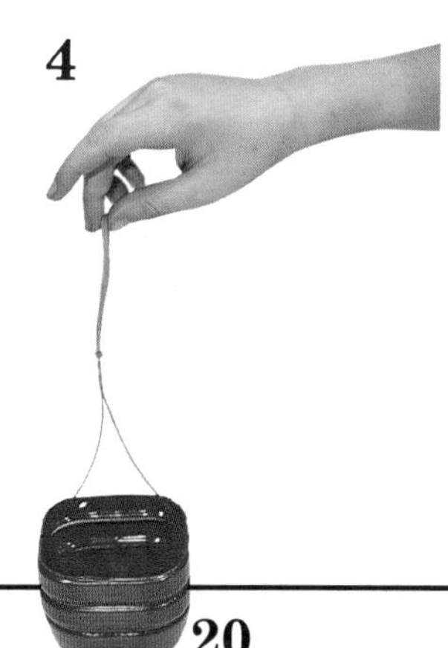

5

5. Position the wood in a secure, vertical position. Attach a pointer to the base of the weighing pan. This will simplify the reading of the scale. Establish the position of the pointer when the pan is empty.

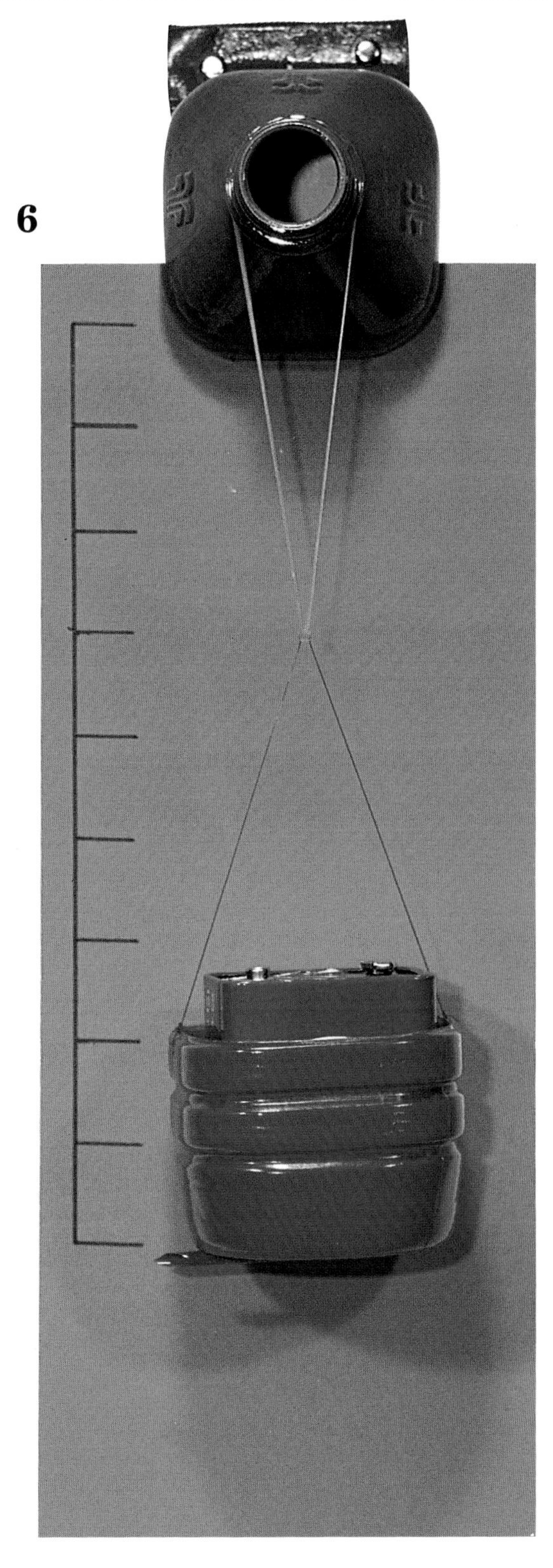

6. Mark the point on the indicating scale where the base of the empty pan rests. All other measurements must be taken from that point. Select a variety of objects to weigh. Place each object inside the weighing pan and read from the scale when the pan is still.

Why It Works

Robert Hooke (1635-1703) devised the theory of elasticity. Hooke's Law states that the extension of a spring is directly proportional to the force applied to it. You observed how much the elastic stretched each time you weighed an object on the "spring" balance. When the weight doubles or trebles, the elastic stretches proportionally. This is because the gravitational pull is stronger for objects containing a greater amount of matter and therefore their weight is greater.

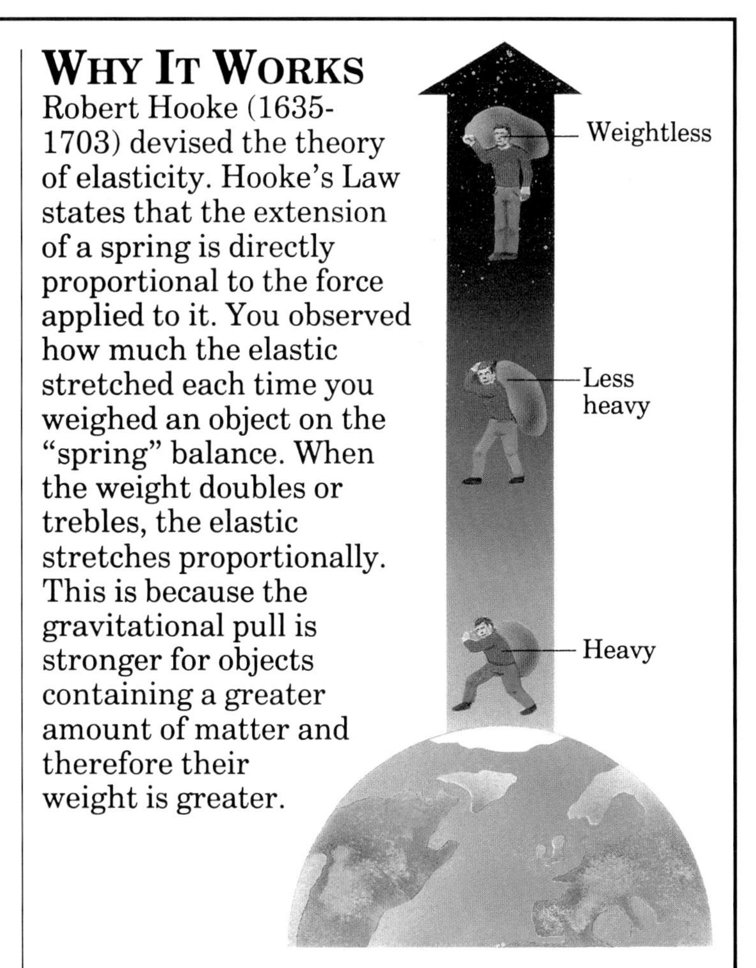

Bright Ideas

Collect a variety of objects to weigh on a real spring balance. Record the length of the spring in mm. Chart these results and look for a relationship between them. The greater the force recorded, the further the spring is likely to stretch. Does the size of an object make a difference to the force of gravity acting on it?

Suspend a stone from a piece of elastic. Now suspend the same object from the elastic and lower it into a bucket of water until it is immersed. What do you observe? Why does this happen?

Tie a long length of string around the middle of a heavy, closed book. Now lift the book by holding the string at each end. Try to make the string horizontal by pulling hard. It is impossible! The force pulling down on the book and the angle created between the two pieces of string makes it very hard to lift.

Measuring Mass

Mass is not the same as weight. Mass is the amount of matter in an object, measured in kilograms – it is constant wherever the object is. Each civilisation developed its own system of 'weight' measurement based on mass. A Roman pound 'weight' had the same mass as one eightieth of a cubic foot of water. Nowadays, weighing has been standardised by the introduction of metric units of mass – kilograms and grammes. A bar of gold has a greater mass than a piece of wood of the same size. Comparing the mass of different materials can be done using a balancing scale.

Balancing Act

1

1. Take two plastic beakers. In the top of each beaker make two holes directly opposite each other. Thread same string through the holes, leaving the same length free on each side.

2

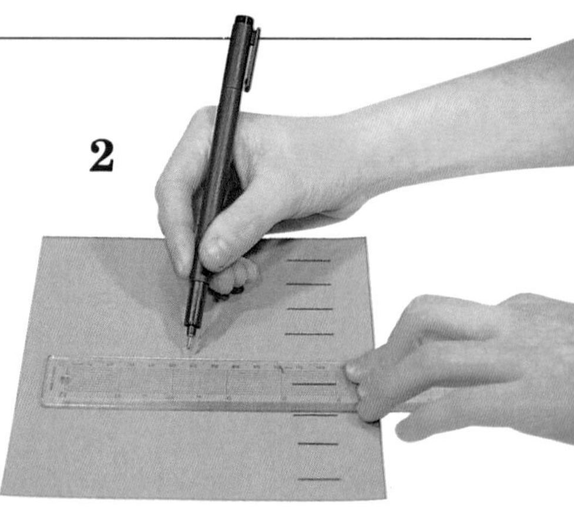

2. On a sheet of card measure a horizontal indicating scale. Be as accurate as possible. The distance between each mark on the scale must be the same. Make the markings clear, so that they can be read easily.

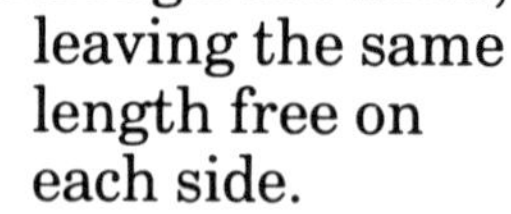

3

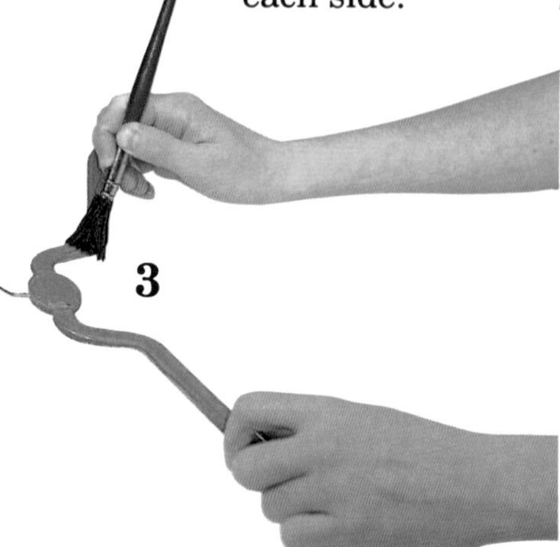

3. Choose a strong coat hanger that has a rigid structure. Decorate the hanger. Cut one end of a plastic straw to make a point. Suspend the straw from the centre of the hanger. Ensure that it hangs vertically.

3

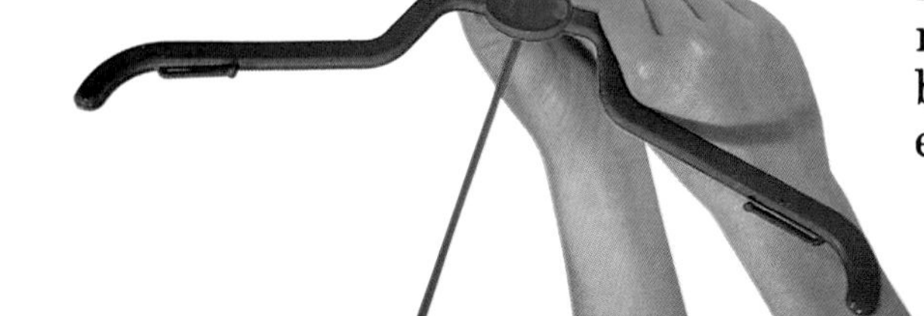

4

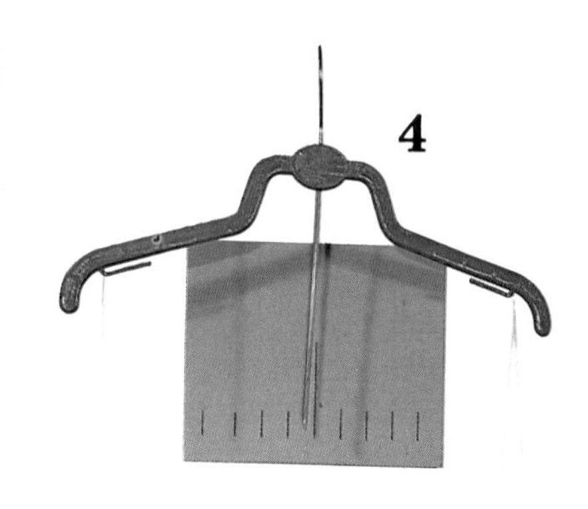

4. Suspend the hanger from a hook. Position the indicating scale so that the straw is in line with the centre mark. Hang both beakers, one from each end of the hanger. Check that they are level.

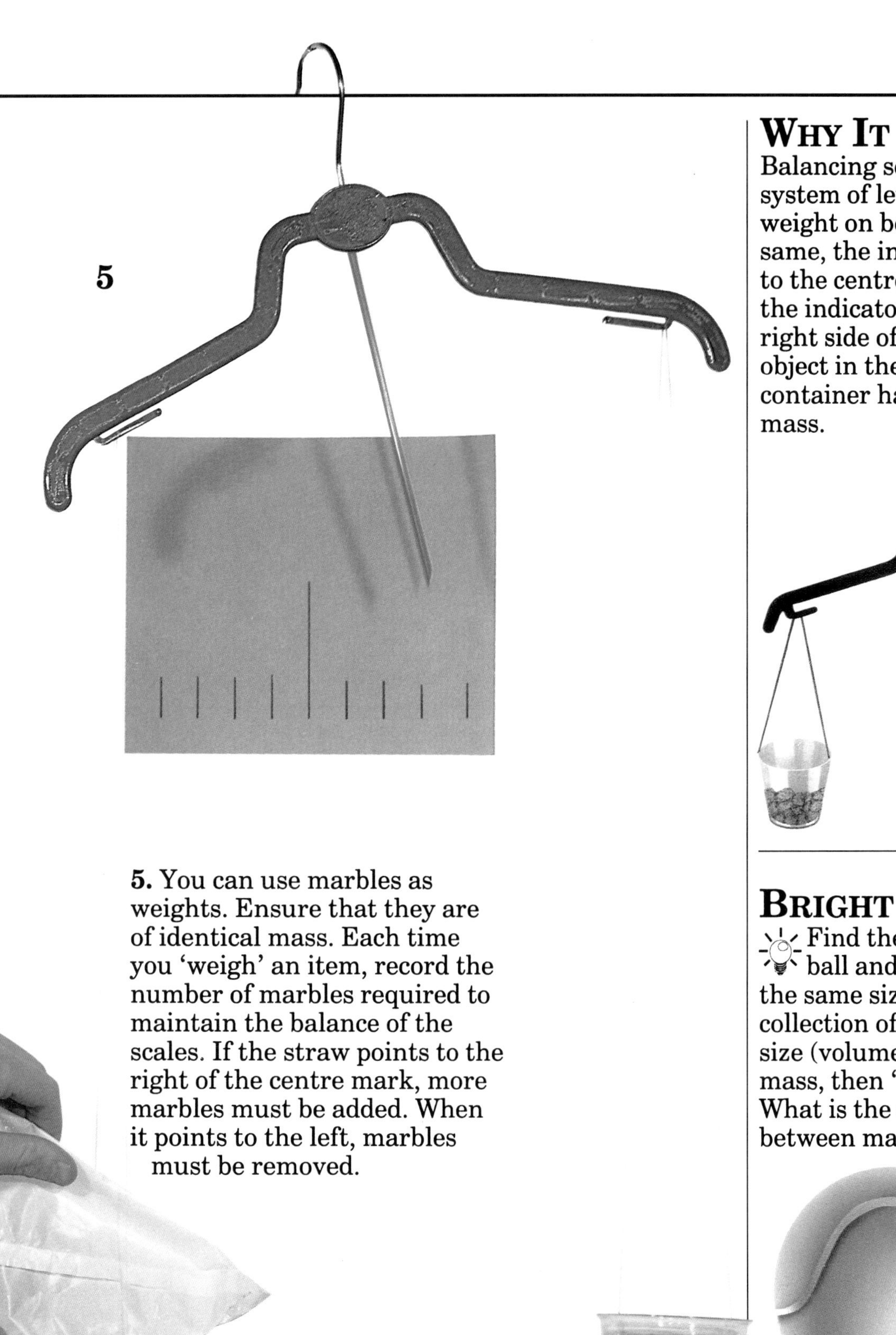

5. You can use marbles as weights. Ensure that they are of identical mass. Each time you 'weigh' an item, record the number of marbles required to maintain the balance of the scales. If the straw points to the right of the centre mark, more marbles must be added. When it points to the left, marbles must be removed.

Why It Works

Balancing scales work on a system of levers. When the weight on both sides is the same, the indicator will point to the centre of the scale. If the indicator points to the right side of the scale, the object in the left-hand container has the greater mass.

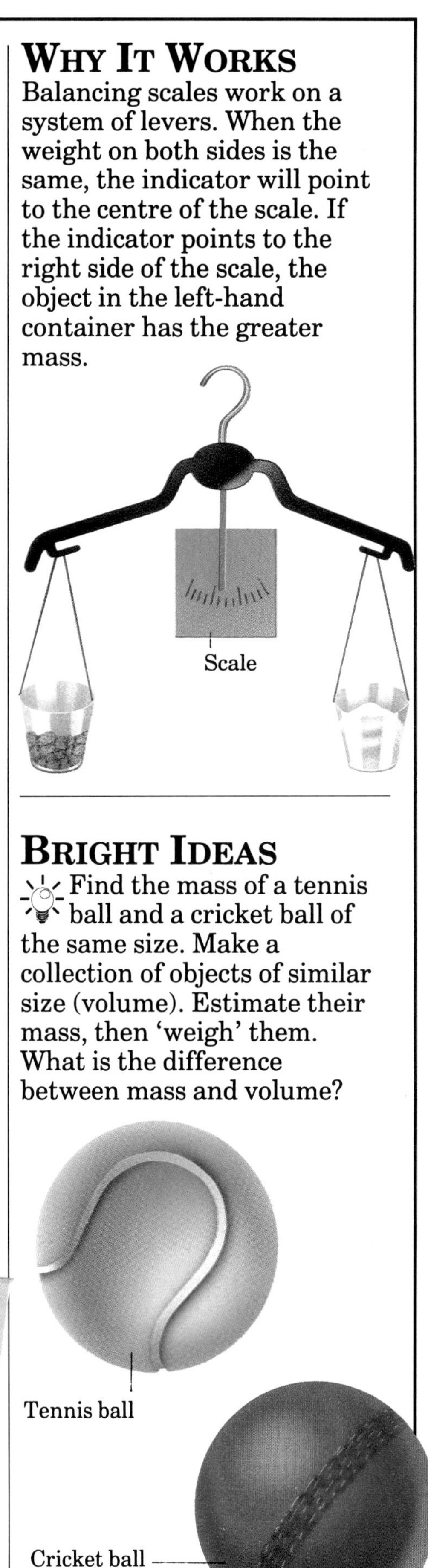

Bright Ideas

Find the mass of a tennis ball and a cricket ball of the same size. Make a collection of objects of similar size (volume). Estimate their mass, then 'weigh' them. What is the difference between mass and volume?

AREA, VOLUME AND CAPACITY

The first crude tools of measurement of length were parts of the body. Shorter units of length were measured against the thumb – the Roman 'uncia' became the inch – the handspan or the foot. Longer lengths were measured against the forearm or a stride. From direct measurements of length, height and depth, indirect measurements of area, volume and capacity can be calculated. The modern unit of measurement for a large area of land is the hectare. A hectare equals 100 acres or 10,000 sq. metres. In 1794, the Cadil was established as the metre standard of capacity – it became known as the litre.

SURFACES AND SPACES

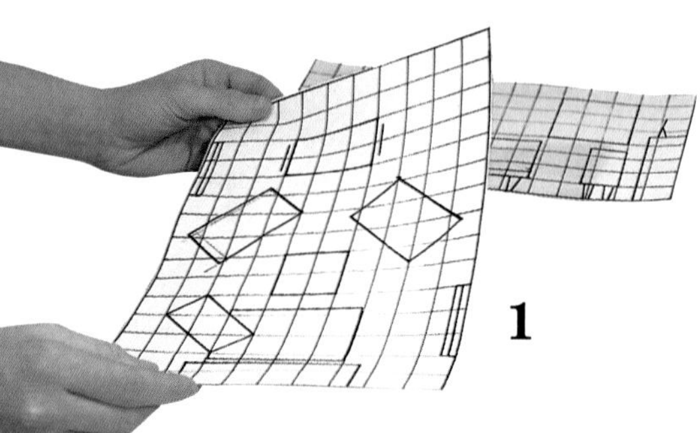

1

1. Use a metre rule or a tape measure to take accurate measurements of a room and the furniture. You need to know all dimensions. Scale down the measurements and transfer them to squared paper.

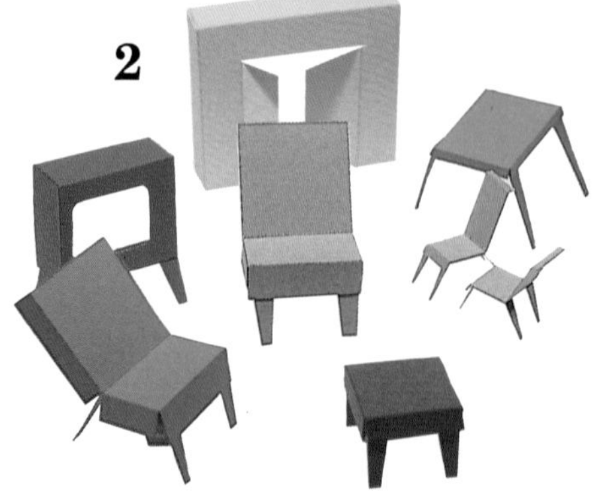

2

2. Each piece of furniture is drawn to scale on squared paper. Use these as patterns to draw round on card. Score along fold lines to achieve a neat 3-dimensionable shape.

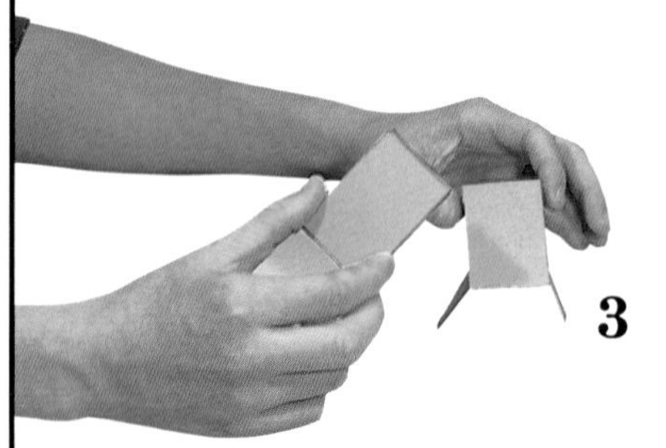

3

3. Construct a scale model of three walls and the floor of the empty room. Draw a net, with flaps, to scale, on squared paper, then transfer it on to card. Cut out the windows and glue together.

4

4. Arrange the furniture inside the 'room', so that it looks like the original room, from which you took measurements.

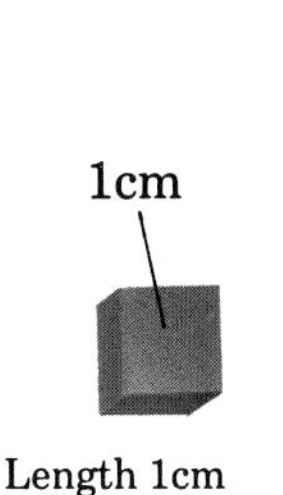

Length 1cm
Volume $1cm^3$

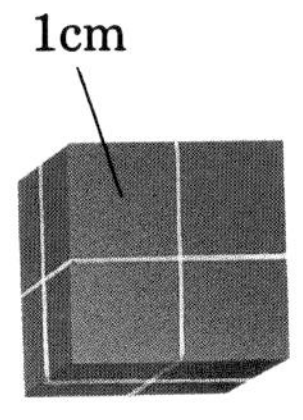

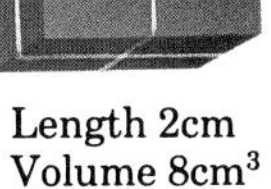

Length 2cm
Volume $8cm^3$

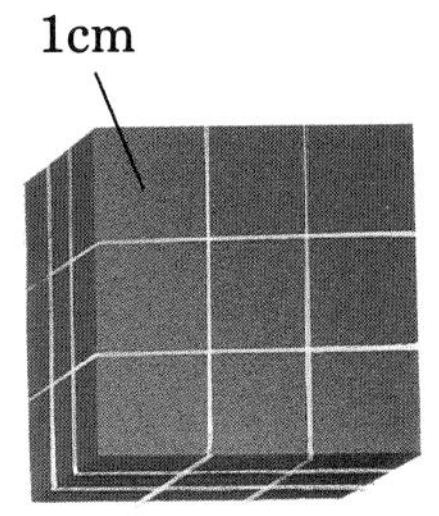

Length 3cm
Volume $27cm^3$

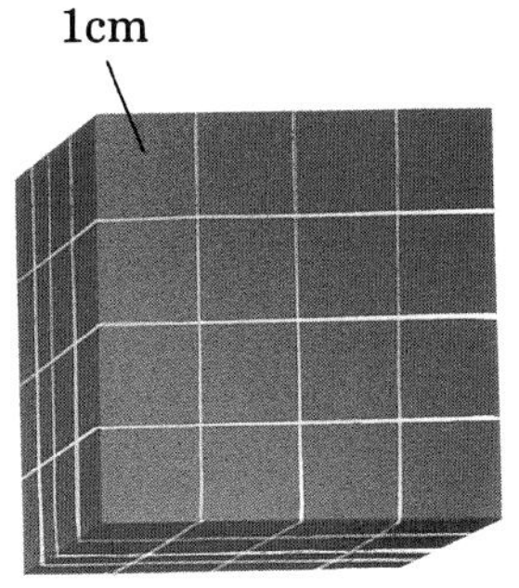

Length 4cm
Volume $64cm^3$

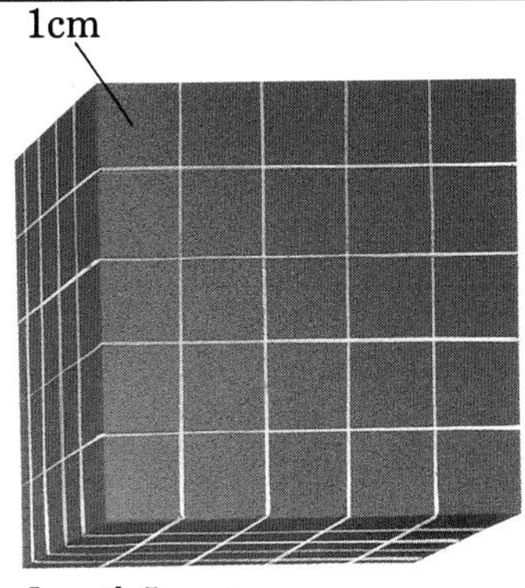

Length 5cm
Volume $125cm^3$

WHY IT WORKS

Area is the size of a surface, it is commonly measured in square units. The area of a rectangular shape is calculated by measuring length times breadth. Volume is the amount of space taken up by a shape. Capacity is the amount, or volume, that a container can hold. All are indirect measurements of size and space. A three-dimensional shape has length, height and depth. The volume of a regular 3-D shape, like a cube, can be measured by finding the product of these three measurements.

5. Observe how much space is taken up by 3-D objects. Be aware of the added dimension of depth. The scale drawings are 2-D, they lack depth. Surface area is flat, space around it can be filled.

BRIGHT IDEAS

Make a tangram puzzle (below) by cutting up a square. See how many shapes you can make using all 7 pieces each time. Will the area be the same each time?

Use 1cm plastic cubes to build a series of larger cubes. How many cubes are used on each face? What is the area of one face? What is the total surface area? What is the volume of each cube? What is the relationship between volume and the surface area of one face?

DENSITY

Britain covers an area of about 244,000 sq. km. France covers an area of approximately 547,000 sq. km. – an area twice as large. A similar number of people live in each. Therefore, France has a lower population density because it has twice as much space for the same number of people. Similarly, the density of a solid or a liquid can only be measured indirectly. It is dependent upon knowing the mass and the volume. Salt water is more dense than fresh water. This is why it is easier to float in the sea, particularly the Dead Sea.

HOW DENSE?

1. Cut a length of straw and, to one end, attach a ball of plasticine. Place it gently in water to ensure that it is balanced.

1

2. First, measure the density of water. Measure out a given amount of water and pour it into a transparent beaker. Place the hydrometer in the water and mark the straw at the surface.

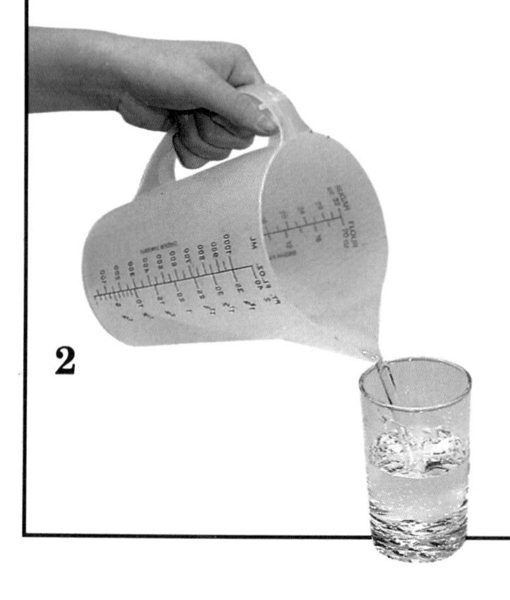

2

Bright Ideas

Collect a variety of objects made from different materials. For example, you could use wood, cork, stone, glass, plastic and metal. Place them in a large container of water. Do they float or sink? An object with a lower density than water will float. Those with a higher density will sink.

Calculate the density of water. Find the mass of an empty beaker. Fill the beaker with a specified volume of water. Find the combined mass of the beaker and the water, then subtract one from the other to find the mass of the water. Use the formula density equals mass divided by volume.

3. Measure out the same quantity of oil and pour it into an identical beaker. Place the hydrometer in the oil and observe the position of the mark on the straw. Repeat for other liquids e.g. vinegar, salt water and sugar water.

Why It Works

You have made an hydrometer to measure the relative density or specific gravity of various liquids. This means that each measurement is compared to the measurement taken for water. The more dense a liquid is, the easier it is for objects to float. In a liquid less dense than water, the straw sinks lower. The lightness or heaviness of a given volume of any substance is called its density. Density is dependent upon how tightly the molecules of each are packed. If various liquids of different densities are gently poured into a transparent container beginning with the most dense first, they will settle in coloured bands according to the density of each. Liquids you can use include salad oil, salt water and coloured fresh water.

Surveying the Land

Tax gatherers in Ancient Egypt were probably the first professional surveyors. They developed a system called triangulation because it was easier to measure angles accurately than it was to measure distances. Surveyors still use the system of triangulation today – they can even use it to measure the height of mountains. The measuring tool used is the theodolite which is mounted on a tripod like the one pictured here. Thales (624-546 BC), a Greek merchant and teacher, devised a method for measuring the height of the Great Pyramid in Ancient Egypt.

What Angle?

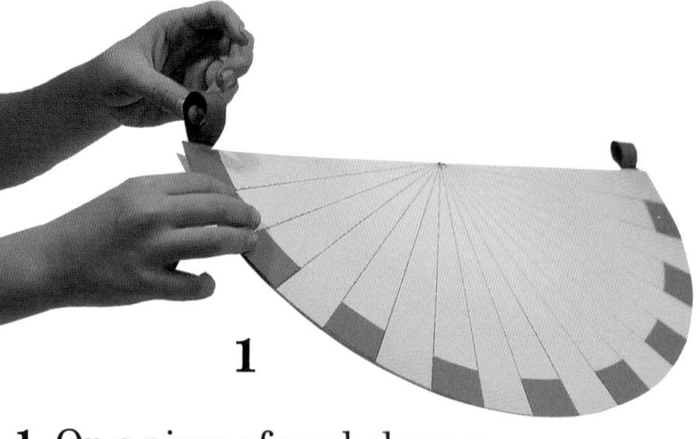

1

1. On a piece of card, draw a circle, radius 30 cm and cut it out with a craft knife. Use a cutting board. Rule a diameter across the circle, then measure and mark divisions of 10 degrees all the way round. There will be 36 divisions, a circle has 360 degrees. Cut the circle in half.

2

2. Mark the divisions clearly. Make two 'viewers' from rolls of card. Secure them between the two semi-circles, at either end, then glue together.

3

3. Wind a strip of card round a heavy object, such as a battery. Attach this to one end of a length of string, at least 60cm long. Secure the other end to the centre of the straight edge of the theodolite.

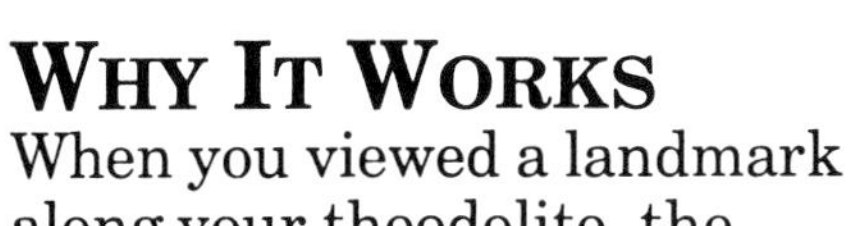

Why It Works

When you viewed a landmark along your theodolite, the plumb line indicated the size of the angle at each end of the base line of an imaginary triangle. Once the distance and the vertical angle is known between two points, a surveyor can calculate the difference in height. That measurement is added to the height of the theodolite to give the total height of the landmark.

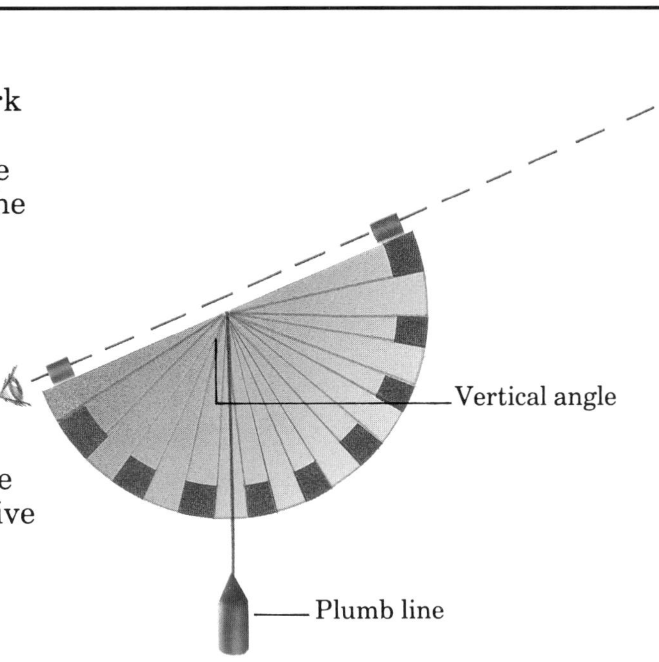

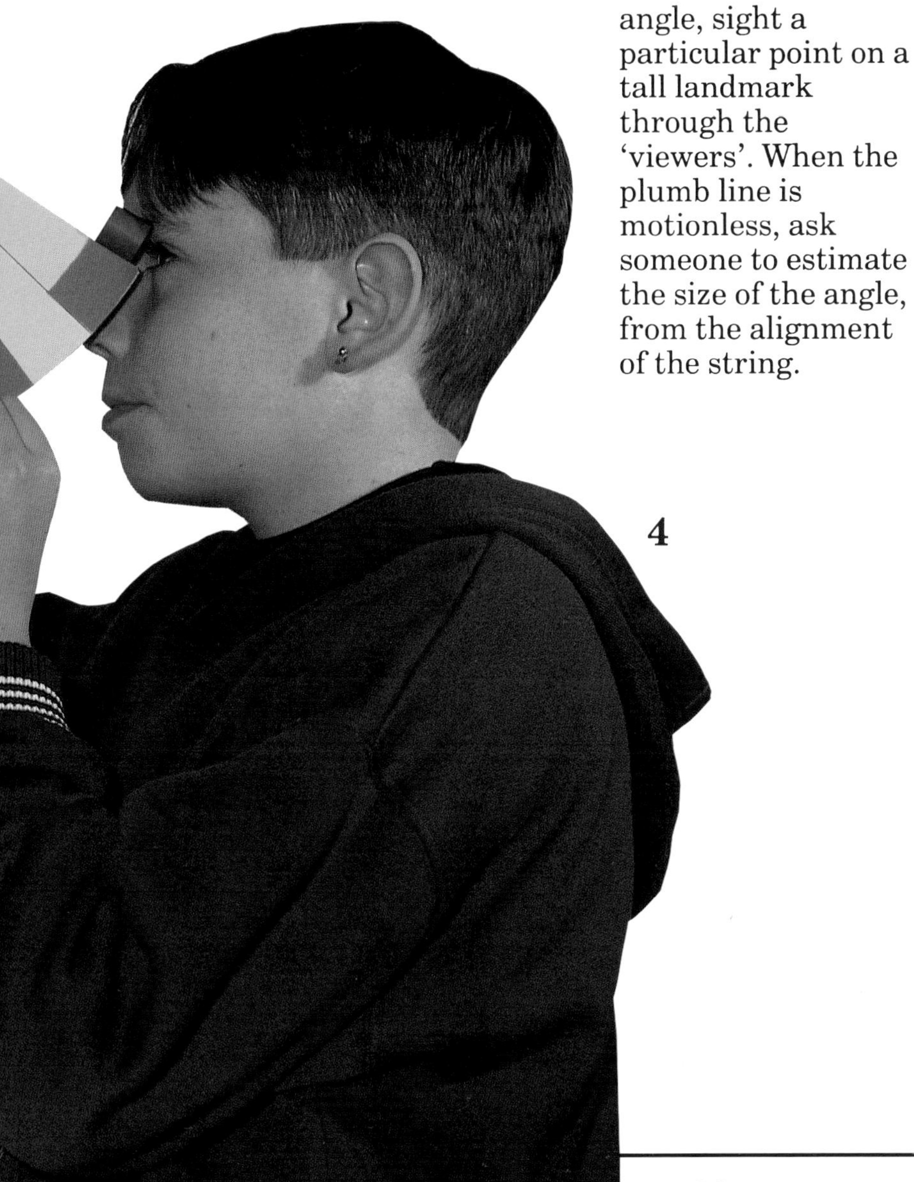

4. To measure an angle, sight a particular point on a tall landmark through the 'viewers'. When the plumb line is motionless, ask someone to estimate the size of the angle, from the alignment of the string.

Bright Ideas

You can make another kind of theodolite out of a sheet of card cut to the shape of a right-angled triangle. You will also need a plumb line, a ruler and a tape measure. By lining up the longest side of the card with the top of a tree or pole, so that the plumb line hangs exactly along the short side nearest to it, you can calculate the height of the tree. You will need to walk towards or away from the tree until this happens. Now measure the distance between yourself and the foot of the tree. Its height is that distance plus your own height.

Measure the height of a friend. Ask them to stand beside a tall tree, then move away until they appear to be 5cm tall on a ruler. Now measure the height of the tree from the same distance. Can you estimate the height of the tree by working out how many times bigger than your friend the tree is?

Reading Weather Signals

Earth has weather because it is surrounded by a shallow blanket of air called the atmosphere. The Sun's rays pass through the atmosphere to reach the ground. The tilt of the Earth on its axis, and the curvature of its surface, create the variety of weather patterns and climatic zones experienced all over the world. For hundreds of years people have tried to forecast the weather. This is done by measuring such things as wind speed using an anemometer like this one, and monitoring any changes. Atmospheric pressure is measured using a barometer. The mercury barometer was invented in 1643 by Torricelli, an assistant of Galileo. The first weather station was built for Napoleon III in 1854.

Stormy Weather

1

1. Take an empty box and remove the lid. Decorate it appropriately with weather symbols.

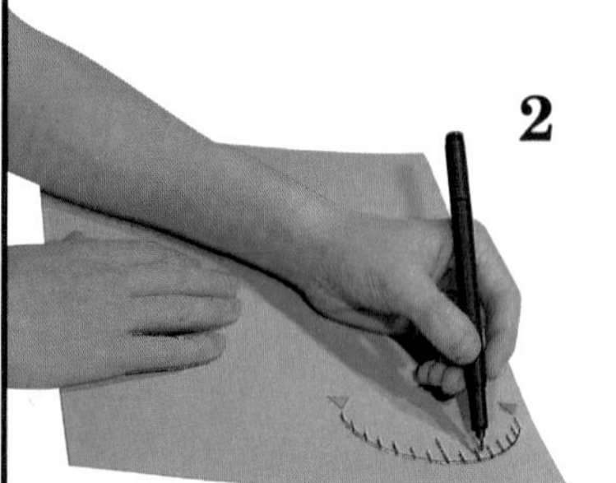

2

2. Measure and draw a curving indicating scale as shown. Make sure the divisions are equal. Use a marker pen to ensure clarity. Indicate 'high' at the top of the scale, and 'low' at the bottom.

3. Cover the box with plastic film, particularly the open top. Ensure that the plastic is taut and crease-free across this open part. This is the 'drum' of your barometer.

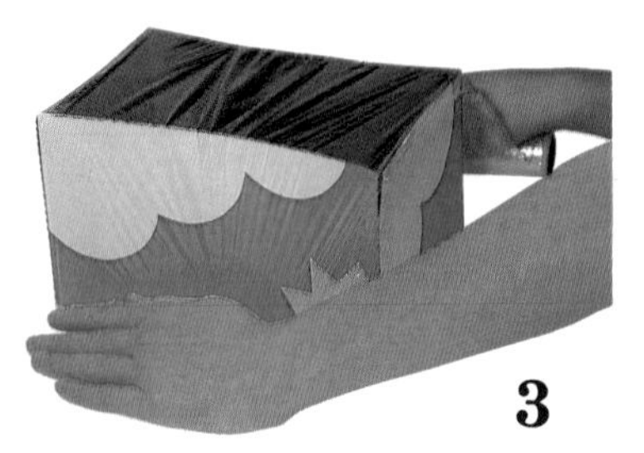

3

4

4. Attach the indicating scale to the back of the box. Make sure that it protrudes beyond the box.

5

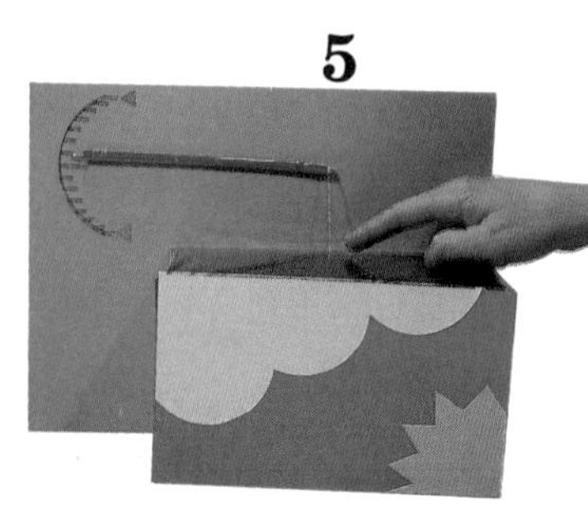

5. Secure the plastic straw horizontally to the scale card with a pin. Direct one end of the straw to the centre of the scale. Fix the other pin inside as a pointer.

WHY IT WORKS

This barometer works as the air pushes down on the cellophane. When air pressure changes, the surface of the cellophane moves up or down. If air pressure falls the pointer falls – rainy weather is approaching. A rise in air pressure means the pointer will rise, indicating dry weather. The anemometer reacts to wind speed. If the wind is strong, it will push the strip of card further up the scale.

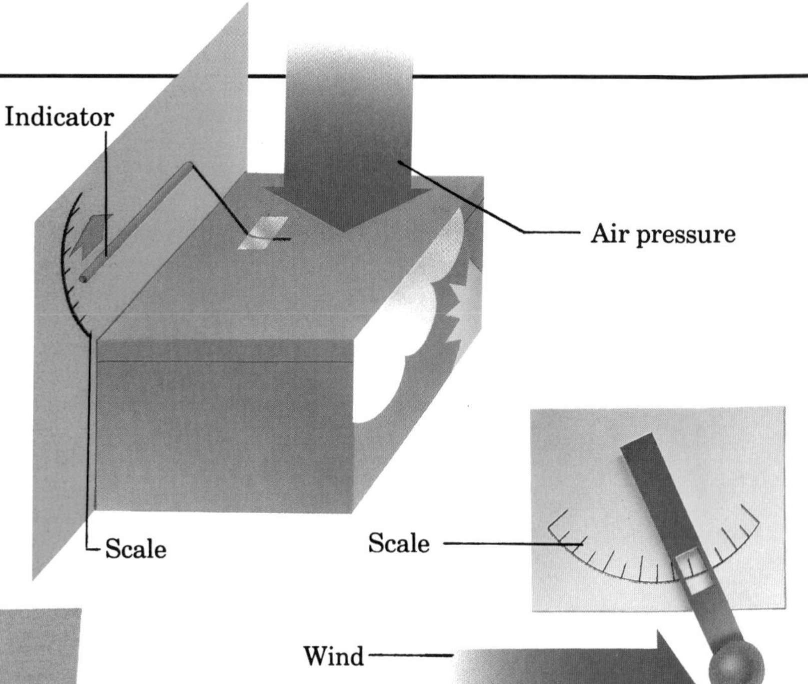

6. Use sticky tape to attach the thread to the straw. Attach the free end of the thread to the surface of the plastic. The straw must be horizontal.

BRIGHT IDEAS

1. Design and make a portable anemometer. Use a protractor and compass to draw a scale on a piece of thick card.

2. Tape half a table tennis ball to a strip of card and cut a window so you can read the scale.

3. Pin the strip to the card. The scale should be:

Angle:	80	60	40	20
km/hr	13	24	34	52

Scientific Terms

ANEROID BAROMETER An instrument used to measure air pressure.
ASTRONOMY The scientific study of the Universe.
CONSTELLATIONS Groups of stars visible in the night sky, recognised easily from their familiar patterns.
FRICTION The force that resists movement when one surface moves relative to another.
GRAVITY The force of attraction between two objects. The more massive an object is, the greater it's force of gravity.
GREENWICH MEAN TIME Standard time throughout most of the world, calculated from the Prime Meridian at Greenwich, London.
IMPERIAL MEASUREMENTS Standard measurements used in Great Britain before the introduction of the Metric System.
LUNAR CYCLE The time taken for the Moon to make one complete revolution of the Earth - approximately 29 days 12 hours 44 minutes 3 seconds.
METEOROLOGIST A person who undertakes the study of weather forecasting.
METRIC SYSTEM The decimal system of units based on the metre.
POLE STAR The star closest to the north celestial pole. It appears to remain fixed in the sky.
SURVEYOR A person who plots detailed maps of the land.
TANGRAM A Chinese puzzle comprising seven cut pieces.
URSA MAJOR A constellation in the Northern hemisphere.

Index